AF508670

Algorithms for Fault Detection and Diagnosis

Algorithms for Fault Detection and Diagnosis

Editors

Francesco Ferracuti
Alessandro Freddi
Andrea Monteriù

MDPI • Basel • Beijing • Wuhan • Barcelona • Belgrade • Manchester • Tokyo • Cluj • Tianjin

Editors

Francesco Ferracuti	Alessandro Freddi	Andrea Monteriù
Università Politecnica delle Marche	Università Politecnica delle Marche	Università Politecnica delle Marche
Italy	Italy	Italy

Editorial Office
MDPI
St. Alban-Anlage 66
4052 Basel, Switzerland

This is a reprint of articles from the Special Issue published online in the open access journal *Algorithms* (ISSN 1999-4893) (available at: https://www.mdpi.com/journal/algorithms/special_issues/Fault_Detection_Diagnosis).

For citation purposes, cite each article independently as indicated on the article page online and as indicated below:

LastName, A.A.; LastName, B.B.; LastName, C.C. Article Title. *Journal Name* **Year**, *Volume Number*, Page Range.

ISBN 978-3-0365-0462-9 (Hbk)
ISBN 978-3-0365-0463-6 (PDF)

Contents

About the Editors

Francesco Ferracuti is an Assistant Professor at Università Politecnica delle Marche (Ancona, Italy), where he teaches "Technology for Automation" and is a founding member of "Revolt srl", a startup that develops data analytics platforms for industrial systems. His research interests include model-based and data-driven fault diagnosis, signal processing, statistical pattern recognition, and their applications in the industry. He has published more than 70 papers in international journals and conferences, and is involved both in national and international research projects.

Alessandro Freddi is an Assistant Professor at Università Politecnica delle Marche (Ancona, Italy), where he teaches "Preventive Maintenance for Robotics and Smart Automation" and is a founding member of "Syncode", a startup operating in the field of industrial automation. His main research activities cover fault diagnosis and fault-tolerant control with applications in robotics and the development and application of assistive technologies. He published more than 80 papers in international journals and conferences, and is involved both in national and international research projects.

Andrea Monteriù is an Associate Professor at Università Politecnica delle Marche (Ancona, Italy). His main research interests include fault diagnosis, fault-tolerant control, nonlinear, dynamics and control, and periodic and stochastic system control, applied in different fields including aerospace, marine, and robotic systems. He has published more than 160 papers in international journals and conferences and is involved both in national and international research projects. He is the author of the book *Fault Detection and Isolation for Multi-Sensor Navigation Systems: Model-Based Methods and Applications* and a co-editor or author of four books on ambient-assisted living including the IET book entitled *Human Monitoring, Smart Health and Assisted Living: Techniques and Technologies*.

Preface to "Algorithms for Fault Detection and Diagnosis"

Due to the increasing demand for security and reliability in manufacturing and mechatronic systems, early detection and accurate diagnosis of faults are key points to reduce the economic losses caused by unscheduled maintenance and downtimes, to increase safety, to prevent the endangerment of human beings involved in the process operations and to improve reliability and availability of autonomous systems.

Mechatronic systems are becoming more heavily digitalized, resulting in more data becoming available for use in detecting and diagnosing faults. This led to a surge of academic effort for developing novel algorithms for systems monitoring.

The development of algorithms for health monitoring and fault and anomaly detection, capable of the early detection, isolation, or even prediction of technical component malfunctioning, is becoming more and more crucial in this context.

This Special Issue is devoted to new research efforts and results concerning recent advances and challenges in the application of "Algorithms for Fault Detection and Diagnosis", articulated over a wide range of sectors. The aim is to provide a collection exploring current state-of-the-art algorithms within this context, together with new advanced theoretical solutions.

The Special Issue includes contributions on the following topics:
- Misalignment fault prediction of wind turbines based on a combined forecasting model
- Genetic algorithm adaptive template matching for offline shape motion tracking based on edge detection
- Blending filter-based thruster fault diagnosis and control recovery evaluation for reusable launch vehicles
- Fault diagnosis of rolling bearing using multiscale amplitude-aware permutation entropy and random forest
- An intelligent warning method for diagnosing underwater structural damage
- A review of lithium-ion battery fault diagnostic algorithms

Francesco Ferracuti, Alessandro Freddi, Andrea Monteriù
Editors

Article

Misalignment Fault Prediction of Wind Turbines Based on Combined Forecasting Model

Yancai Xiao * and Zhe Hua

School of Mechanical, Electronic and Control Engineering, Beijing Jiaotong University, Beijing 100044, China; 17121266@bjtu.edu.cn
* Correspondence: ycxiao@bjtu.edu.cn; Tel.: +86-010-51684273

Received: 20 January 2020; Accepted: 27 February 2020; Published: 1 March 2020

Abstract: Due to the harsh working environment of wind turbines, various types of faults are prone to occur during long-term operation. Misalignment faults between the gearbox and the generator are one of the latent common faults for doubly-fed wind turbines. Compared with other faults like gears and bearings, the prediction research of misalignment faults for wind turbines is relatively few. How to accurately predict its developing trend has always been a difficulty. In this paper, a combined forecasting model is proposed for misalignment fault prediction of wind turbines based on vibration and current signals. In the modelling, the improved Multivariate Grey Model (IMGM) is used to predict the deterministic trend and the Least Squares Support Vector Machine (LSSVM) optimized by quantum genetic algorithm (QGA) is adopted to predict the stochastic trend of the fault index separately, and another LSSVM optimized by QGA is used as a non-linear combiner. Multiple information of time-domain, frequency-domain and time-frequency domain of the wind turbine's vibration or current signals are extracted as the input vectors of the combined forecasting model and the kurtosis index is regarded as the output. The simulation results show that the proposed combined model has higher prediction accuracy than the single forecasting models.

Keywords: misalignment; fault prediction; combined prediction; multivariate grey model; quantum genetic algorithm; least squares support vector machine

---

1. Introduction

The problem of energy shortage and environmental degradation in the world is becoming more and more serious. Wind energy as environmentally friendly and renewable energy has attracted increasing attention [1]. The cumulative installed capacity of global wind power has also steadily increased in recent years [2]. Because wind turbines are often located in remote areas and the working environment is poor, many wind turbines often fail during operation, which greatly reduces their work quality and efficiency and increases maintenance costs [3]. Therefore, how to effectively decrease the risk of fault during the operation of wind turbines has become a difficult problem.

At present, doubly-fed wind turbines (DFWT) have become the main units for large-capacity wind farms [4]. Due to installation errors, deformation after loading or frequent wind speed fluctuations, misalignment between the gearbox and the generator often happens [5]. The misalignment fault of wind turbines belongs to a latent fault [6,7]. This is because when it happens in actual operation, the unit's operating parameters will not reach their early warning values immediately, but when the fault accumulates to a certain extent, it will seriously damage the unit's equipment and cause unit shutdown [8]. Therefore, it is necessary to predict the latent trend of misalignment, which can overcome the blindness of handling the fault and avoid more loss of human and material resources.

There are many commonly used signals for mechanical fault prediction, such as vibration signals, current signals, temperature signals, pressure signals, etc. [9–13]. When the equipment fails, the amplitude of the mechanical vibration will increase [14]. Therefore, vibration signals often more quickly and directly reflect the operational status of the equipment. Compared with vibration signals, the current signals can be more easily obtained and are not easily affected by noise. Thus, in this paper, the vibration signal and current signal are regarded as the signal sources to research the misalignment fault of wind turbines.

After the fault signals are obtained, a reasonable and effective prediction method is necessary for accurately predicting the future operational status of the equipment faults. At present, many scholars have studied the prediction techniques for different types of faults [15–17]. In order to determine a suitable forecasting model of misalignment fault in wind turbines, the commonly used prediction methods are summarized in Table 1.

Table 1. Summary of common prediction methods.

Method	Scope	Advantage	Disadvantage
Time series prediction	Stationary random sequence Short-term forecast	Few samples demand Simple model	Not suitable for non-linear systems Not suitable for medium to long-term forecasting
Support Vector Machine	Small sample Non-linear system	High prediction accuracy Relatively few samples	The selection of parameters has an impact on the accuracy of the model
Least Squares Support Vector Machine	Small sample Non-linear system	Predictive calculation speed is higher than SVM Suitable for few samples	The selection of parameters has an impact on the accuracy of the model
Neural networks	Non-linear complex system	Strong nonlinear mapping ability Adaptive learning ability	Local minimum problem A lot of samples required
Grey prediction model	Data with specific trends Short-term forecast	Few samples High modelling accuracy	Incomplete consideration Not suitable for long-term forecasting

For the complex non-linear system, a single forecasting model is not enough to obtain ideal prediction results. Therefore, in order to predict the mechanical fault accurately, the combined forecasting model has attracted more and more attention from scholars. For example, in Ref. [18], the improved Grey Model (GM (1,1)) and the Back Propagation (BP) neural network optimized by Genetic Algorithm (GA) were used as the single forecasting models. The minimum sum of error squares was used as the combination principle to assign appropriate weight coefficients to them. The combined forecasting model had a smaller prediction error. Ref. [19] proposed a calculation method of combined weight coefficients for the unequal weight of error. The combined forecasting model was constructed by Multivariate Grey Model (MGM $(1, n)$) and Extreme Learning Machine (ELM) neural network. The combined forecasting model was more suitable for predicting the trend of the bearing fault. In Ref. [20], according to the minimum variance principle, Support Vector Machine (SVM) and grey model were combined to make up the shortcomings of single forecasting models. In Ref. [21], SVM was used as the combiner of forecasting models. The Kalman filter, BP neural network and SVM model were used as single forecasting models. The prediction errors of the single forecasting models were larger than that of the combined model. In Ref. [22], the BP neural network was used to determine the weight coefficients of each single forecasting model. The combined forecasting model using GM $(1,1,\theta)$ optimized by Particle Swarm Optimization (PSO) algorithm and SVM optimized by PSO achieved better prediction accuracy for the short-term load of a regional power grid.

Because the wind turbine is a complex non-linear system, when the misalignment fault occurs, the fault signals have both certainty and randomness characteristics [23]. In addition, the misalignment fault samples obtained in this paper are relatively few. It can be indicated from Table 1 that the grey prediction model is suitable for predicting deterministic trends with few samples, while Least Squares Support Vector Machine (LSSVM) is suitable for predicting the non-linear and stochastic trends with few samples and higher speed [24]. Therefore, the grey prediction model and LSSVM are selected to be the prediction methods for the misalignment fault of wind turbines. In the grey prediction models, the MGM (1, n) can use the multivariate characteristic parameters of the fault state as the inputs of the prediction model [25], which can comprehensively reflect the fault state at the previous moment and establish a more accurate prediction model. Therefore, this paper uses MGM (1, n) to predict the misalignment faults of wind turbines. Because MGM (1, n) has the disadvantage of only being suitable for short-term prediction, the rolling prediction method is used to improve MGM (1, n). Compared with the combination of fixed weight coefficient, LSSVM can assign non-linear weight coefficients to the prediction values of single prediction models dynamically, which makes the weight allocated more reasonably. Therefore, in this paper, a combined forecasting model using LSSVM optimized by quantum genetic algorithm (QGA) as a non-linear and variable weight combiner is proposed. The output prediction values of both the LSSVM model optimized by QGA and the improved Multivariate Grey Model (IMGM (1, n)) are input to the non-linear combiner to get the final predicted values. The vibration and current signals from the misalignment fault simulation model of wind turbine demonstrate that the combined forecasting model has higher prediction accuracy than the single ones.

2. Forecasting Method and Application

2.1. Multivariate Grey Model

In 1982, Professor Julong Deng of Huazhong University of Science and Technology proposed the grey system theory, which has good adaptability to small samples and uncertain systems [26]. Because the GM (1,1) model only uses single time series data and it cannot reflect the interaction between multiple variables, some scholars have proposed a MGM (1, n), where n is the number of variables. The MGM (1, n) is not a simple combination of GM (1,1), but a generalization from the univariate GM (1,1) to the multivariate case by solving n differential equations simultaneously [27].

It is assumed that $x_i^{(0)}(k), (i = 1, 2, \ldots, n; k = 1, 2, \ldots, m)$ is n sets of time series data, and k is the number of each data set.

The process of establishing a MGM (1, n) prediction model is as follows:

(1) Accumulate the original data to generate a new sequence data $x_i^{(1)}(k)$:

$$x_i^{(1)}(k) = \sum_{j=1}^{k} x_i^{(0)}(j) \tag{1}$$

(2) A system of n-ary first-order ordinary differential equations can be used to express MGM (1, n):

$$\begin{cases} \frac{dx_1^{(1)}}{dt} = a_{11}x_1^{(1)} + a_{12}x_2^{(1)} + \cdots + a_{1n}x_n^{(1)} + b_1 \\ \frac{dx_2^{(1)}}{dt} = a_{21}x_1^{(1)} + a_{22}x_2^{(1)} + \cdots + a_{2n}x_n^{(1)} + b_2 \\ \quad\quad\quad\quad \vdots \\ \frac{dx_n^{(1)}}{dt} = a_{n1}x_1^{(1)} + a_{n2}x_2^{(1)} + \cdots + a_{nn}x_n^{(1)} + b_n \end{cases} \tag{2}$$

Equation (2) can be written in matrix form as:

$$\frac{dX^{(1)}(t)}{dt} = AX^{(1)}(t) + B \tag{3}$$

$$\text{where, } A = \begin{bmatrix} a_{11} & a_{12} & \cdots & a_{1n} \\ a_{21} & a_{22} & \cdots & a_{2n} \\ \vdots & \vdots & \vdots & \vdots \\ a_{n1} & a_{n2} & \cdots & a_{nn} \end{bmatrix}, X^{(1)}(t) = \begin{bmatrix} x_1^{(1)}(k) \\ x_2^{(1)}(k) \\ \vdots \\ x_n^{(1)}(k) \end{bmatrix}, B = \begin{bmatrix} b_1 \\ b_2 \\ \vdots \\ b_n \end{bmatrix}.$$

(3) By the method of least squares, the parameter matrices A and B can be estimated. Assume that $a_i = [a_{i1}, a_{i2}, \ldots, a_{in}, b_i]^{\mathrm{T}}, (i = 1, 2, \ldots, n)$. The corresponding matrix can be expressed as:

$$\hat{a}_i = (L^{\mathrm{T}} L)^{-1} L^{\mathrm{T}} Y_i \tag{4}$$

where, $Y_i = [x_i^{(0)}(2), x_i^{(0)}(3), \ldots, x_i^{(0)}(m)]^{\mathrm{T}}$, $L = (L_1, L_2, \ldots, L_j, \ldots, L_n, I)$,

$$L_j = \begin{bmatrix} (x_j^{(1)}(2) + x_j^{(1)}(1))/2 \\ (x_j^{(1)}(3) + x_j^{(1)}(2))/2 \\ \vdots \\ (x_j^{(1)}(m) + x_j^{(1)}(m-1))/2 \end{bmatrix}$$ and I is the unit matrix.

(4) The predicted values of MGM $(1, n)$ can be obtained.

$$\hat{X}^{(1)}(k) = e^{\hat{A}(k-1)} X^{(1)}(1) + \hat{A}^{-1}(e^{\hat{A}(k-1)} - I) \cdot \hat{B}, (k = 1, 2, \ldots) \tag{5}$$

$$\hat{X}^{(0)}(1) = X^{(0)}(1) \tag{6}$$

$$\hat{X}^{(0)}(k) = \hat{X}^{(1)}(k) - \hat{X}^{(1)}(k-1), (k = 2, 3, \ldots) \tag{7}$$

When $n = 1$, MGM $(1, n)$ model is transformed into GM $(1,1)$ model.

2.2. Improved Multivariate Grey Model

Compared with GM $(1, 1)$, MGM $(1, n)$ has the advantage of considering multiple input features at the same time. However, both the GM $(1,1)$ and MGM $(1, n)$ are not suitable for medium to long-term forecasting. Because the rolling prediction method can make the prediction model achieve medium to long-term forecasting, MGM $(1, n)$ is combined with the rolling prediction method in this paper to get the improved MGM $(1, n)$ prediction model (IMGM $(1, n)$). The schematic diagram of IMGM $(1, n)$ is shown in Figure 1.

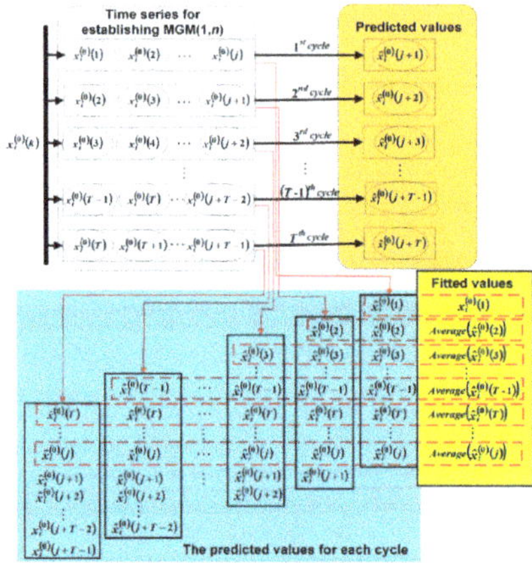

Figure 1. Schematic diagram of the improved Multivariate Grey Model (IMGM $(1, n)$).

In the IMGM (1, n), based on the establishing process of MGM (1, n), the rolling update of the modelling data is performed by adding the actual value corresponding to the previous step of the current prediction point and removing the first one in the previous modelling data to achieve the dynamic addition of new information. The modelling data is updated every time the model outputs one predicted value.

Assume that $x_i^{(0)}(k), (i = 1, 2, \ldots, n; k = 1, 2, \ldots, m)$ is n sets of sequence data. The number of each data set input to the MGM (1, n) is j, which is $x_i^{(0)}(k), (i = 1, 2, \ldots, n; k = 1, 2, \ldots, j$ and $j \leq m)$. The number of cycles is T and the number of predicted steps for each cycle is one. The predicted process of IMGM (1, n) is as follows:

- As shown in Figure 1, the original data $\{x_i^{(0)}(1), x_i^{(0)}(2), \ldots, x_i^{(0)}(j)\}, (i = 1, 2, \ldots, n)$ are used to establish the MGM (1, n). The model outputs one predicted value and j fitted values. The oldest data $x_i^{(0)}(1)$ is removed and the actual data $x_i^{(0)}(j + 1)$ is added to reconstruct the MGM (1, n). The above steps are cycled T times.
- When T cycles are finished, T predicted values after $x_i^{(0)}(j)$ are obtained, and the final j fitted values of each data set output from IMGM (1, n) are the average of the fitted values in the corresponding order for all cycles.

Thus, the IMGM (1, n) is suitable for medium-long prediction. In the application of this paper, the number j of each data set input to the MGM (1, n) model is 45. The number of T is 15. When the 15 cycles are finished, the 45 averages of the fitted values and 15 predicted values will be obtained.

2.3. LSSVM Optimized by Quantum Genetic Algorithm

The Least Squares Support Vector Machine (LSSVM) replaces inequality constraints with equality constraints, regarding the sum of squared errors as experience losses of training set, transforming quadratic programming problems into linear equations [28]. The Radial Basis Function (RBF) is simple to calculate, requiring few parameters to be determined, and with strong generalization ability. The equation of the Radial Basis Function (RBF) is as follows:

$$K(x_i, x_j) = \exp\left(-\frac{\|x_i - x_j\|^2}{2\sigma^2}\right) \tag{8}$$

where σ is the kernel width. In this paper, the RBF kernel function is used as the kernel function of LSSVM. The regularization parameter γ and the parameter σ^2 of the RBF kernel function have a great influence on the prediction accuracy of LSSVM. If the value of γ is too large or too small, the generalization ability of the system will be deteriorated. The value of σ^2 will also affect the performance of the model. Therefore, choosing appropriate parameters can give LSSVM a good prediction effect. In this paper, the QGA is used to realize the adaptive selection of the parameters of LSSVM.

The QGA is an intelligent optimization algorithm combining quantum computing and genetic algorithms, which was proposed by K. H. Han et al. [29]. The probability amplitude representation of qubits is applied to the coding of the chromosome so that a chromosome can express the superposition of multiple states. The operation of quantum revolving gate can update the chromosome, and thus the optimal solution of the goal can be achieved [30]. Compared with GA, QGA has the characteristics of small population size, strong optimization ability, high convergence speed and short calculation time [31].

In GA, the commonly used coding methods for chromosomes are binary coding, real coding and symbol coding. In QGA, chromosomes are encoded using qubits [32].

The state of a qubit can be expressed as:

$$|\varphi\rangle = \alpha|0\rangle + \beta|1\rangle \tag{9}$$

where, α and β are complex constants and satisfy $|\alpha|^2 + |\beta|^2 = 1$. Qubits are used to store and express a gene. The gene can be in a "0" state or a "1" state, or any superposition state of them, which makes QGA have better diversity characteristics than GA. Multi-state genes encoded by qubits are shown in Equation (10).

$$q_j^t = \left[\begin{array}{c|c|c|c|c|c|c|c} \alpha_{1,1}^t & \alpha_{1,2}^t & \cdots & \alpha_{1,k}^t & \cdots & \alpha_{m,1}^t & \alpha_{m,2}^t & \cdots & \alpha_{m,k}^t \\ \beta_{1,1}^t & \beta_{1,2}^t & \cdots & \beta_{1,k}^t & \cdots & \beta_{m,1}^t & \beta_{m,2}^t & \cdots & \beta_{m,k}^t \end{array} \right] \tag{10}$$

where, q_j^t is the jth chromosome of the tth generation; $\alpha_{i,j}^t$ and $\beta_{i,j}^t$, $(i = 1, 2, \ldots, m; j = 1, 2, \ldots, k)$ contain the quantization information of the gene. k is the number of qubits of each gene and m is the number of chromosome genes.

For QGA, the quantum revolving gate in quantum theory is used to achieve population update. The operations of selection, crossover and mutation of GA are not adopted. The matrix of quantum revolving gate is expressed as Eequation (11):

$$\boldsymbol{U}(\theta_i) = \left[\begin{array}{cc} \cos(\theta_i) & -\sin(\theta_i) \\ \sin(\theta_i) & \cos(\theta_i) \end{array} \right] \tag{11}$$

The update process is as follows:

$$\left[\begin{array}{c} \alpha_i' \\ \beta_i' \end{array} \right] = \left[\begin{array}{cc} \cos(\theta_i) & -\sin(\theta_i) \\ \sin(\theta_i) & \cos(\theta_i) \end{array} \right] \left[\begin{array}{c} \alpha_i \\ \beta_i \end{array} \right] \tag{12}$$

where, $[\alpha_i\ \beta_i]^{\text{T}}$ is the ith qubit of the chromosome. $[\alpha_i'\ \beta_i']^{\text{T}}$ is the new qubit after the update. θ_i is the rotation angle and it is determined according to a previously designed adjustment strategy [33].

Before using QGA to optimize the parameters of LSSVM, the fitness function needs to be defined first. In this paper, the Root Mean Square Error (RMSE) is used as the objective function. The calculation formula of RMSE is shown in Equation (13):

$$\min f(\gamma, \sigma^2) = \sqrt{\frac{1}{n} \sum_{i=1}^{n} (y_i - \hat{y}_i)^2} \tag{13}$$

where, y_i is the actual value of the ith sample; $\hat{y}_i$ is the predicted value of the ith sample; $\gamma \in [\gamma_{\min}, \gamma_{\max}]$, $\sigma^2 \in [\sigma^2_{\min}, \sigma^2_{\max}]$, $i = 1, 2, \ldots, n$. The idea of parameter optimization of LSSVM model is to search for a set of parameters $[\gamma, \sigma^2]$ through the iterative of QGA to minimize the objective function. The flow chart of QGA is shown in Figure 2. In this paper, the searching range of the regularization parameter γ is set to [0.1, 100], the searching range of the parameter σ^2 of kernel function is [0.01, 100], the maximum number of iterations is 200 and the population size is 20, and the coded length of the quantum chromosome is 20.

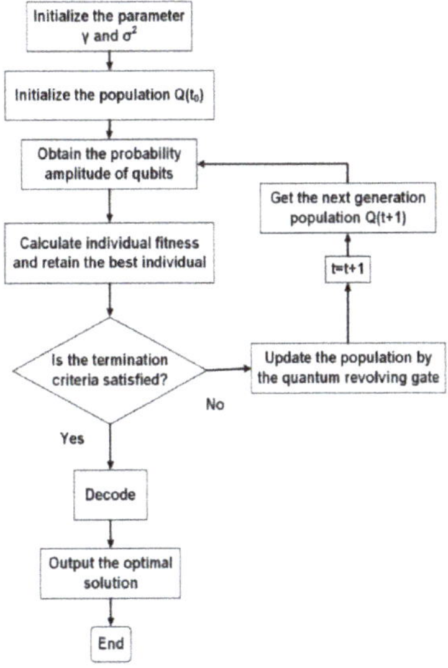

Figure 2. The flow chart of the Least Squares Support Vector Machine (LSSVM) optimized by quantum genetic algorithm (QGA).

2.4. Combined Prediction

The combined prediction was first proposed by Bates. J. M and Granger. C. W. J. The combined forecasting model is constructed by assigning different weights to the prediction values of single forecasting models [34]. The classification of combined prediction can be divided into the following two types:

- According to the functional relationship between the combined and the single forecasting models, the combined forecasting model can be divided into linear and non-linear combination prediction [35].
- According to the weight coefficients of the single models, the combined forecasting model can be divided into fixed weight and variable weight combination prediction [36].

The linear combination prediction has relatively large errors and has great limitations, while the fixed weight combination prediction cannot dynamically adjust the combination weight, therefore it is necessary to use a non-linear and variable weight combined forecasting model. For example, neural networks and SVM are non-linear combiners, and these two combination methods can realize non-linear and variable weight combination of the single forecasting models. However, neural networks are not suitable for processing few samples data. There will be overfitting problems and the prediction accuracy is not satisfactory [37]. SVM or LSSVM has obvious advantages in solving few samples, non-linear and high-dimensional problems [38]. Therefore, LSSVM optimized by QGA is adopted in this paper as a non-linear and variable weight combiner. The flow chart of the proposed combined prediction is shown in Figure 3.

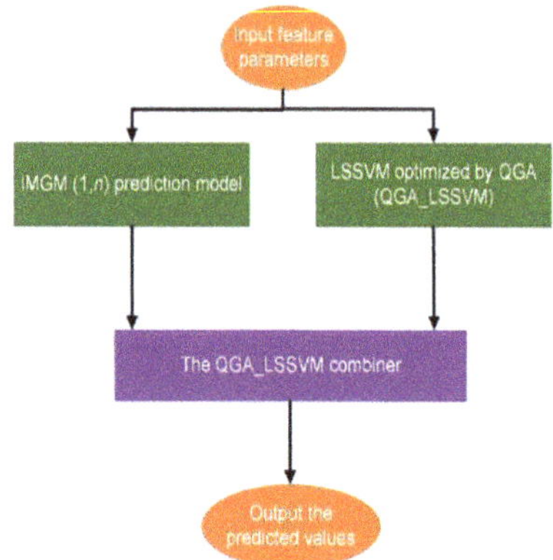

Figure 3. The flow chart of combined prediction.

2.5. Evaluation Criteria for Forecasting Models

2.5.1. Accuracy Test of Grey Prediction Model

After the grey prediction model is established, the model can be used to predict effectively only after the accuracy test is qualified. The posteriori error test is adopted in this paper. The equations of posteriori error test are as follows.

The posteriori error ratio C:

$$C = \frac{S_1}{S_2} \tag{14}$$

where, S_1 is the standard deviation of the original sequence and S_2 is the standard deviation of the residual sequence.

The small error probability P:

$$P = P\left\{\left|\varepsilon(i) - \bar{\varepsilon}\right| < 0.6745 S_1\right\}, (i = 1, 2, \ldots, k) \tag{15}$$

where, $\varepsilon(i)$ is the residual error and $\bar{\varepsilon}$ is the residual average.

The smaller the value of C and the greater the value of P, the higher the grade of the grey prediction model. Generally, the accuracy grade of the grey prediction model can be divided into four levels which are shown in Table 2.

Table 2. The precision grade of accuracy test.

Precision Grade	C	P
Level 1 (Good)	$C \leq 0.35$	$0.95 \leq P$
Level 2 (Qualified)	$0.35 < C \leq 0.5$	$0.80 \leq P < 0.95$
Level 3 (Barely qualified)	$0.5 < C \leq 0.65$	$0.70 \leq P < 0.80$
Level 4 (Unqualified)	$0.65 < C$	$P < 0.70$

Where the precision grade of grey prediction model is the maximum of the grade of P and the grade of C.

2.5.2. Forecasting Evaluation Index

In order to compare the prediction effects of the combined model and the single ones, the following 3 indexes are calculated.

Assume that $y = (y_1, y_2, \ldots, y_N)$ is the actual value of the data, $\hat{y} = (\hat{y}_1, \hat{y}_2, \ldots, \hat{y}_N)$ is the predicted value, and $\bar{y}$ is the average of the actual value.

Root Mean Square Error (RMSE):

$$RMSE = \sqrt{\frac{1}{N}\sum_{i=1}^{N}(y_i - \hat{y}_i)^2}, \; RMSE \in [0, +\infty) \tag{16}$$

The closer the RMSE is to zero, the smaller the prediction error and vice versa.
Mean Absolute Error (MAE):

$$MAE = \frac{1}{N}\sum_{i=1}^{N}|y_i - \hat{y}_i|, \; MAE \in [0, +\infty) \tag{17}$$

The closer the MAE is to zero, the smaller the prediction error and vice versa.
Coefficient of determination (R^2):

$$R^2 = 1 - \frac{\sum\limits_{i=1}^{N}(y_i - \hat{y}_i)^2}{\sum\limits_{i=1}^{N}(y_i - \overline{y})^2}, \; R^2 \in [0, 1] \tag{18}$$

The larger the R^2 is and the closer it is to 1, the better the fitting effect of the prediction model and vice versa [39].

3. Signal Acquisition and Feature Extraction

3.1. Signal Acquisition

A 1.5 MW wind turbine is taken as the research object in this paper. The three-dimensional model of the 1.5 MW wind turbine drive system is established by Solidworks, and then it is imported into ADAMS for dynamic simulation. In the dynamic simulation model, parallel misalignment, angular misalignment and comprehensive misalignment are simulated. The acceleration signal of the high-speed output of the gearbox is used as the vibration signal (details in literature [40]). In MATLAB, the wind turbine and its control system are established to obtain the stator current for misalignment faults (details in literature [41]). The simulation model of the wind turbine is in the Maximum Power Points Tracking (MPPT) stage, in which the input speed is 81.3°/s, and the vibration signal and stator current signal are obtained under parallel misalignment fault for researching.

3.2. Feature Extraction

3.2.1. Time-Domain Feature Parameters

Time-domain analysis is used to directly obtain the time-domain statistical parameters of fault signals in the time domain. It can simply and intuitively represent the changes of the fault signals. The time-domain feature parameters include dimensional parameters and dimensionless parameters. Assume that the signal obtained is a discrete sequence $x_i, (i = 1, 2, 3, \dots, N)$ with finite length, where N represents the number of data points of each sample. Three commonly used time-domain feature parameters are used in the paper. The equations are as follows.
Root Mean Square (RMS):

$$X_{RMS} = \sqrt{\frac{1}{N}\sum_{i=1}^{N}x_i^2} \tag{19}$$

The RMS represents the average energy measure of the signal and is very stable [42].

Kurtosis:

$$K = \frac{1}{N} \sum_{i=1}^{N} x_i^4 \tag{20}$$

The kurtosis is highly sensitive to early faults and impact signals, but its stability is poor [43,44]. Kurtosis index:

$$K_f = \frac{K}{X_{RMS}^4} \tag{21}$$

The kurtosis index is very sensitive to the impact of vibration signals [45].

3.2.2. Frequency-Domain Feature Parameters

Time-domain analysis can only briefly and roughly represent the fault of the device, but it cannot find the inherent cause of the fault. Frequency-domain analysis can transform the signal from the time-domain to the frequency-domain through Fourier transform. Different frequency components are clearly displayed to find the fault frequency. Assume that x_i, $(i = 1, 2, 3, \ldots, N)$ is a finite-length discrete time series and the sampling frequency is f_c. The calculation equation of commonly used frequency-domain feature parameters are as follows.

Mean Square Frequency (MSF):

$$MSF = \frac{1}{4\pi^2 f_c^2} \cdot \frac{\int_0^\pi \omega^2 S(\omega) d\omega}{\int_0^\pi S(\omega) d\omega} \tag{22}$$

Centroid Frequency (FC):

$$FC = \frac{1}{2\pi f_c} \cdot \frac{\int_0^\pi \omega S(\omega) d\omega}{\int_0^\pi S(\omega) d\omega} \tag{23}$$

Variance Frequency (VF):

$$VF = \frac{1}{4\pi^2 f_c^2} \cdot \frac{\int_0^\pi (\omega - 2\pi f_c FC) S(\omega) d\omega}{\int_0^\pi S(\omega) d\omega} = MSF - FC^2 \tag{24}$$

where $S(\omega)$ is the power spectrum of the signal. $S(\omega) = X(\omega) \cdot \overline{X(\omega)}$, $X(\omega) = \sum_{i=1}^{N} x_i e^{-j\pi\omega}$. ω is the discretized angular frequency.

3.2.3. Time-Frequency Domain Feature Parameters

The Empirical Mode Decomposition (EMD) improved by mirror extension is used to process the vibration signal [46]. The advantage of this method is the suppression of endpoint effects. The energy entropy, which is suitable for non-stationary and non-linear complex signals [47], is extracted from the processed vibration signal.

Assume [48]:

$$E_i = \int \left| c_i(t) \right|^2 dt = \sum_{k=1}^{n} \left| x_{ik}^2 \right|, \tag{25}$$

$$E = \sum_{i=1}^{n} E_i, \tag{26}$$

$$p_i = \frac{E_i}{E}, \tag{27}$$

The equation of energy entropy P_i is as follows:

$$P_i = -\sum_{i=1}^{n} p_i \log_{10} p_i \tag{28}$$

where, $i = 1, 2, 3 \ldots n$; x_{ik} is the amplitude of each discrete point; E_i is the energy of each Intrinsic Mode Function (IMF) component, and E is the total energy of the signal. After vibration signals at different times being decomposed by IEMD, the number of IMF may be different. Therefore, the correlation coefficient method is used to determine the effective number of IMF components. Assume that the non-stationary signal $x(t)$ is decomposed into a finite number of uncorrelated components $x_i(t)$ by IEMD. The correlation coefficient between each IMF and the original signal can be calculated as:

$$\rho_{x_i x} = \frac{\sum\limits_{t=0}^{\infty} x_i(t)x(t)}{\left[\sum\limits_{t=0}^{\infty} x_i^2(t) \sum\limits_{t=0}^{\infty} x^2(t)\right]^2} \tag{29}$$

where, $i = 1, 2, \ldots, n$ and $0 \le |\rho_{x_i x}| \le 1$. The larger the correlation coefficient, the greater the degree of correlation and vice versa. When $|\rho_{x_i x}| > 0.1$, it usually indicates that the IMF has a good correlation with the original signal [49].

The stator current signal obtained from the simulation model is decomposed into 4 layers by the Dual-tree Complex Wavelet Transform (DTCWT) [50]. The obtained 5 sub-band signals are reconstructed and the sample entropy is calculated. Sample entropy can obtain better and more stable results with less data and has stronger robustness [51]. For finite sequences N_t, the equation for the sample entropy Se_i is as follows:

$$Se(m, r, N_t) = -\ln \frac{\overline{B}^{m+1}(r)}{\overline{B}^{m}(r)} \tag{30}$$

where, m is the number of dimensions of the constructed vector, r is a given threshold, and $\overline{B}^{m}(r)$ is the average of the maximum distance between two m-dimensional vectors.

3.3. Feature Vectors and Normalization

The input speed of the dynamic simulation model of the wind turbine drive system is set to be 81.3°/s. Afterwards, 60 segments of vibration and current signals of parallel misalignment faults are collected. The time-domain, frequency-domain and time-frequency domain feature parameters of each signal segment are extracted to construct feature vectors. After extracting the time-frequency domain feature parameters of the vibration signal, the correlation coefficients of the first three IMFs of the vibration signal decomposed by IEMD are all greater than 0.1. Therefore, the first three IMFs are selected as effective components in this paper. The feature parameters of vibration and current signal are listed in Tables 3 and 4.

Table 3. The indexes of feature vector for vibration signal.

Feature Vector	Feature	Index
Vibration features library (9 dimensions)	Time domain	root mean square, kurtosis, kurtosis index
	Frequency domain	mean square frequency, center of gravity frequency, frequency variance
	Time-frequency domain	energy entropy of the first 3 IMF components of IEMD

Table 4. The indexes of feature vector for current signal.

Feature Vector	Feature	Index
Current features library (11 dimensions)	Time-domain	root mean square, kurtosis, kurtosis index
	Frequency-domain	mean square frequency, center of gravity frequency, frequency variance
	Time-frequency domain	5 sample entropy obtained by 4-layer decomposition of DTCWT

The fault feature parameters have different orders of magnitude. In order to facilitate the processing of data and ensure that the program runs faster when it converges, the Mapminmax function is used to normalize the time-domain, frequency-domain and time-frequency domain feature parameters in this paper. The equation of Mapminmax function is as follows:

$$y = \frac{y_{max} - y_{min}}{x_{max} - x_{min}} \cdot (x - x_{min}) + y_{min} \tag{31}$$

where, x is an unnormalized matrix, y_{min} is a minimum value of the normalization interval, and y_{max} is a maximum value of the normalization interval. The feature parameters are normalized to the range [0, 1] in this paper.

For the vibration and current signals, multivariate feature parameters with 9 dimensions and 11 dimensions are used as input vectors in this paper. Because the kurtosis is sensitive to early faults [43,44], the kurtosis is selected as the output of the prediction model.

4. Fault Prediction Results and Discussion

4.1. Fault Prediction Results and Discussion of Vibration Signals

4.1.1. The Results of IMGM (1, n)

The MGM (1, n) is suitable for the prediction of few samples and the required number of modelling is usually between 5 and 60 [52–54]. Therefore, the collected 60 vibration signal samples are divided into the first 45 samples and the last 15 samples according to the ratio of 3:1. The first 45 samples are used to construct an IMGM (1, n) prediction model. The indexes of the posteriori error test calculated from the fitted values are listed in Table 5.

Table 5. The precision grade of the Multivariate Grey Model (MGM (1,9)) and IMGM (1,9) for vibration signal.

Method	C	P	Precision Grade
MGM (1,9)	0.4952	0.8667	Qualified
IMGM (1,9)	0.4880	0.9048	Qualified

It can be shown from Table 5 that the models of MGM (1,9) and IMGM (1,9) belong to the qualification grade, so they can be used for prediction. Because the C value of IMGM (1,9) is smaller and the P value of IMGM (1,9) is larger, IMGM (1,9) is more suitable for predicting the kurtosis of vibration signals than MGM (1,9). Figure 4 shows the prediction results of MGM (1,9) and IMGM (1,9) of vibration signals.

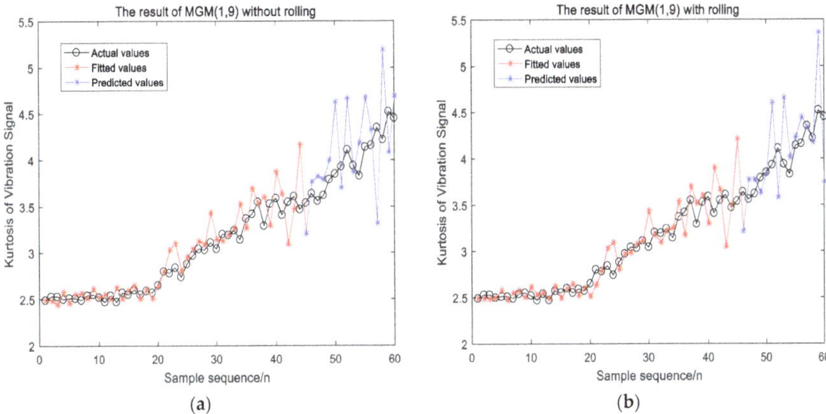

Figure 4. The result of MGM (1,9) and IMGM (1,9) for vibration signal. (**a**) The result of MGM (1,9); (**b**) the result of IMGM (1,9).

In Figure 4, the predicted and fitted values of MGM (1,9) and IMGM (1,9) both fluctuated up and down around their actual values. The forecasting evaluation indexes of IMGM (1,9) and MGM (1,9) are listed in Table 6. For the fitted values, the RMSE and R^2 of IMGM (1,9) is nearly the same to that of MGM (1,9), while the MAE of IMGM (1,9) is better than that of MGM (1,9). For the predicted values, the RMSE, MAE, and running time of the IMGM (1,9) are smaller than those of MGM (1,9). The R^2 of IMGM (1,9) is larger than that of MGM (1,9). Therefore, IMGM (1,9) improves the prediction accuracy of MGM (1,9) for misalignment fault vibration signal of wind turbine. This is because when predicting the last 15 points, the data used to establish the MGM (1,9) model has always been the first 45 points, and the data has not been updated with the increase prediction steps. However, the modelling data of IMGM (1,9) is updated by adding the actual value corresponding to the previous step of the current prediction point. The modelling data is updated every time the model output one predicted value. Hence, the IMGM (1,9) can be one of the single prediction models input to the combined forecasting model for vibration signal.

Table 6. Comparison of forecasting evaluation indexes of MGM (1,9) and IMGM (1,9) for the vibration signal.

Method	Date Set	RMSE	MAE	R^2	Time(s)
MGM (1,9)	Fitted values	0.1946	0.1313	0.9071	1.8776
	Predicted values	**0.5047**	**0.4086**	**0.3392**	
IMGM (1,9)	Fitted values	0.1959	0.1232	0.9053	1.6780
	Predicted values	**0.4349**	**0.3358**	**0.5673**	

4.1.2. The Results of LSSVM Optimized by QGA

The LSSVM optimized by QGA (QGA_LSSVM) is also adopted to predict the kurtosis of misalignment fault. The same first 45 samples are used to train the LSSVM model. The other 15 samples are used as the testing set. The results are compared with the LSSVM optimized by GA (GA_LSSVM) and the LSSVM optimized by Grid Search (Grid Search_LSSVM). The parameters $[\gamma, \sigma^2]$ obtained by the three prediction models are listed in Table 7.

Table 7. The optimized parameters for the vibration signal.

Method	γ	σ^2
QGA_LSSVM	100	99.9658
GA_LSSVM	55.8399	58.4184
Grid Search_LSSVM	89.5265	28.8621

The three forecasting models are established based on the optimized parameters in Table 7. The corresponding forecasting evaluation indexes are listed in Table 8.

Table 8. The forecasting evaluation indexes of the vibration signal.

Method	Date Set	RMSE	MAE	R^2	Time(s)
Grid Search_LSSVM	Training set	0.0095	0.0078	0.9992	2.6590
	Testing set	**0.1550**	**0.1296**	**0.3687**	
GA_LSSVM	Training set	0.0158	0.0126	0.9979	3.2358
	Testing set	**0.1422**	**0.1182**	**0.5274**	
QGA_LSSVM	Training set	0.0148	0.0118	0.9981	1.8286
	Testing set	**0.1175**	**0.0975**	**0.7229**	

In Table 8, for the training set, the indexes of Grid Search_LSSVM show that it has smaller prediction errors of training set than GA_LSSVM and QGA_LSSVM. For the testing set, the RMSE and MAE of QGA_LSSVM are smaller than those of GA_LSSVM and Grid Search_LSSVM. The R^2 of QGA_LSSVM is the largest and its running time is the shortest. Therefore, compared with the results of Grid Search_LSSVM and GA_LSSVM, QGA_LSSVM has the best prediction accuracy and the shortest running time, so it can be one of the single forecasting models input to the combined forecasting model for vibration signal.

4.1.3. The Results of the Combined Forecasting Model

In this paper, the prediction results of IMGM (1,9) and QGA_LSSVM are used as the inputs of QGA_LSSVM combiner and the actual kurtosis values of the vibration signal are used as the output. The comparison of predicted results between the single forecasting models and combined model are shown in Figure 5.

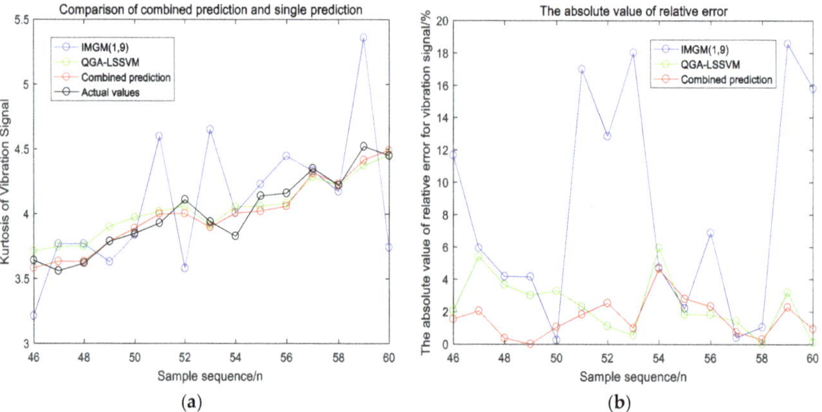

Figure 5. Comparison of the combined forecasting model and the single forecasting models for vibration signal. (**a**) The prediction result from the 46th points to the 60th points; (**b**) the absolute value of relative error.

In Figure 5a, the fluctuation range of IMGM (1,9) around the actual value is the largest. The prediction results of QGA_LSSVM and the combined forecasting model are closer to the actual value. Figure 5b shows that the absolute value of relative error of IMGM (1,9) is significantly large. From the 46th predicted points to the 51th predicted points, the absolute value of relative error of the combined forecasting model is less than that of QGA_LSSVM. After the 53th predicted point, they have similar absolute value of relative error. Therefore, the combined forecasting model has a smaller absolute value of relative error than the two single forecasting models. It can also be shown from Table 9 that, for the training test, the forecasting evaluation indexes of the combined forecasting model are near to those of QGA_LSSVM and better than those of IMGM (1,9); but for the testing set, the indexes of the combined forecasting model are the best in Table 9. Thus, compared with the single forecasting models, the combined forecasting model improves the prediction accuracy and reduces the running time for the vibration signal.

Table 9. The forecasting evaluation indexes of the vibration signal.

Method	Data Set	RMSE	MAE	R^2	Time(s)
Combined prediction	Training Set	0.0179	0.0150	0.9973	1.6025
	Testing Set	**0.0840**	**0.0688**	**0.9100**	
QGA_LSSVM	Training Set	0.0148	0.0118	0.9981	1.8286
	Testing Set	**0.1175**	**0.0975**	**0.7229**	
IMGM (1,9)	Fitted values	0.1959	0.1232	0.9053	1.6780
	Predicted values	**0.4349**	**0.3358**	**0.5673**	

For vibration signal, the optimal parameters of the combiner are $[\gamma, \sigma^2] = [24.5174, 9.4170]$.

4.2. Fault Prediction Results and Discussion of Current Signal

4.2.1. The Results of IMGM $(1, n)$

The collected 60 current signal samples are also divided into the first 45 samples and the last 15 samples according to the ratio of 3:1. The first 45 samples are also used to construct an IMGM $(1, n)$ prediction model. Eleven feature parameters of the current signal are used as inputs and the kurtosis is used as output. The indexes of the posteriori error test of current signal are listed in Table 10.

Table 10. The precision grade of MGM (1,11) and IMGM (1,11) for the current signal.

Method	C	P	Precision Grade
MGM (1,11)	0.6378	0.7632	Barely qualified
IMGM (1,11)	0.4002	0.8857	Qualified

In Table 10, MGM (1,11) is barely qualified grade and IMGM (1,11) belongs to the qualified grade. Therefore, IMGM (1,11) is more suitable for predicting the kurtosis of the current signal than MGM (1,11). The prediction results of MGM (1,11) and IMGM (1,11) of the current signal are shown in Figure 6.

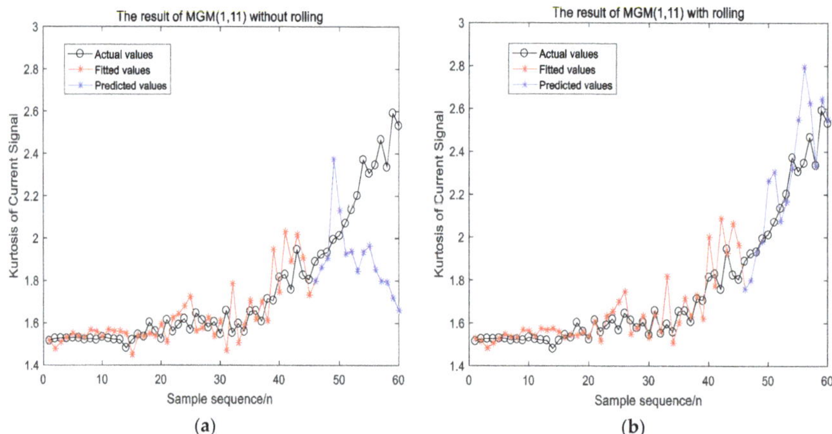

Figure 6. The result of MGM (1,11) and IMGM (1,11) for the current signal. (**a**) The result of MGM (1,11); (**b**) the result of IMGM (1,11).

By comparing the prediction results in Figure 6a,b, the predicted fluctuation range of IMGM (1,11) is obviously smaller than that of MGM (1,11). It is because when predicting the last 15 points, the data used to establish the MGM (1,11) has always been the first 45 points, and the data has not been updated with the increase prediction steps. However, the modelling data of IMGM (1,11) is updated by adding the actual value corresponding to the previous step of the current prediction point. The modelling data is updated every time the model outputs one predicted value. Hence, when predicting the kurtosis of a current signal with a large rising slope, the IMGM (1,11) has significantly better prediction accuracy than that of MGM (1,11). The forecasting evaluation indexes of IMGM (1,11) and MGM (1,11) are listed in Table 11. For the fitted values, the RMSE of IMGM (1,11) is nearly the same to that of MGM (1,11), while the MAE and R^2 of IMGM (1,11) in the table is better than those of MGM (1,11). For the predicted values, the RMSE, MAE and running time of the IMGM (1,11) are smaller than those of MGM (1,11). The R^2 of IMGM (1,11) is much larger than that of MGM (1,11). Therefore, compared with MGM (1,11), IMGM (1,11) improves the prediction accuracy and stability of the current signal, and it can be one of the single prediction models input to the combined forecasting model for current signal.

Table 11. Comparison of forecasting evaluation indexes of MGM (1,11) and IMGM (1,11) for the current signal.

Method	Data Set	RMSE	MAE	R^2	Time(s)
MGM (1,11)	Fitted values	0.0898	0.0690	0.6267	1.9489
	Predicted values	**0.4594**	**0.3732**	**0.2725**	
IMGM (1,11)	Fitted values	0.0893	0.0563	0.7265	1.8974
	Predicted values	**0.1723**	**0.1216**	**0.7546**	

4.2.2. The Results of LSSVM Optimized by QGA

For the current signal, Grid Search_LSSVM, GA_LSSVM and QGA_LSSVM are also used for prediction. The same first 45 current samples are also used to train the LSSVM. The other 15 samples are also used as the testing set. The optimized parameters $[\gamma, \sigma^2]$ obtained by the three prediction models are listed in Table 12.

Table 12. The optimized parameters for the current signal.

Method	γ	σ^2
QGA_LSSVM	100	100
GA_LSSVM	26.2473	53.0292
Grid Search_LSSVM	108.9959	30.3934

The three forecasting models are established based on the optimized parameters in Table 12. The corresponding forecasting evaluation indexes are listed in Table 13.

Table 13. The forecasting evaluation indexes of the current signal.

Method	Data Set	RMSE	MAE	R^2	Time(s)
Grid Search_LSSVM	Training set	0.0056	0.0043	0.9994	2.6170
	Testing set	**0.1283**	**0.0978**	**0.6103**	
GA_LSSVM	Training set	0.0169	0.0129	0.9940	2.6152
	Testing set	**0.1119**	**0.0854**	**0.7415**	
QGA_LSSVM	Training set	0.0101	0.0073	0.9979	1.8338
	Testing set	**0.0777**	**0.0582**	**0.9054**	

For the training set, the Grid Search_LSSVM has smaller errors than GA_LSSVM and QGA_LSSVM. For the testing set, all of forecasting evaluation indexes of QGA_LSSVM are better than those of GA_LSSVM and Grid Search_LSSVM in Table 13. Thus, QGA_LSSVM has a good prediction effect not only for the vibration signal but also for the current signal, and QGA_LSSVM can be one of the single forecasting models input to the combined forecasting model for the current signal.

4.2.3. The Results of the Combined Forecasting Model

The prediction results of IMGM (1,11) and QGA_LSSVM are used as the inputs of the QGA_LSSVM combiner, and the actual kurtosis values of the current signal are used as the output. The predicted results are shown in Figure 7.

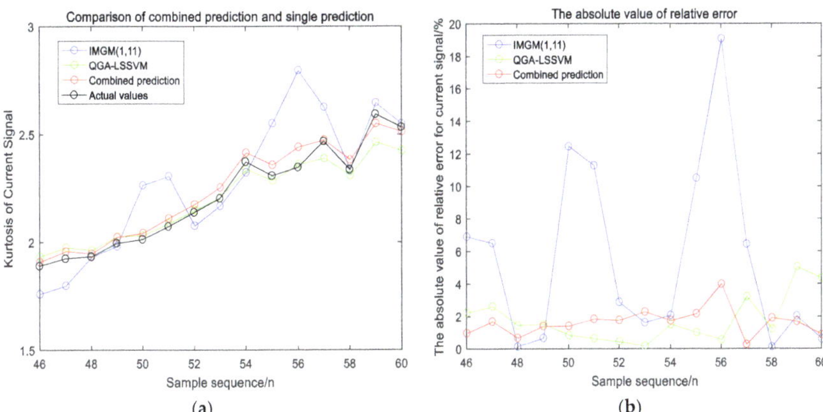

Figure 7. Comparison of the combined forecasting model and the single forecasting models for current signal. (**a**) The prediction result from the 46th points to the 60th points; (**b**) the absolute value of relative error.

Figure 7a indicates that, compared with IMGM (1,11), the prediction results of QGA_LSSVM and the combined forecasting model are closer to the actual values. In Figure 7b, the absolute value of relative error of QGA_LSSVM and combined forecasting model is within 6%. The absolute value of relative error of IMGM (1,11) is much larger than those of other models. In Table 14, for the training set, the forecasting evaluation indexes of the combined forecasting model are near to those of QGA_LSSVM and better than those of IMGM (1,11), but for the testing set, the indexes of the combined forecasting model are the best. Therefore, it is proved by current signals that the combined forecasting model has a higher prediction accuracy with less running time.

Table 14. The forecasting evaluation indexes of the current signal.

Method	Data Set	RMSE	MAE	R^2	Time(s)
Combined Prediction	Training Set	0.0162	0.0116	0.9944	1.5843
	Testing Set	**0.0592**	**0.0520**	**0.9635**	
QGA_LSSVM	Training Set	0.0101	0.0073	0.9979	1.8338
	Testing Set	**0.0777**	**0.0582**	**0.9054**	
IMGM (1,11)	Fitted values	0.0893	0.0563	0.7265	1.8974
	Predicted values	**0.1723**	**0.1216**	**0.7546**	

For the current signal, the optimal parameters of the combiner are $[\gamma, \sigma^2] = [35.2214, 100]$.

5. Conclusions

To improve the prediction accuracy of single forecasting models, this paper proposes a combined forecasting model using LSSVM optimized by QGA as a combiner to predict the misalignment fault of wind turbines based on vibration and current signals. The specific method is to extract the time-domain, frequency-domain and time-frequency domain feature parameters from the signals at first. After being normalized, the extracted features are used to predict by IMGM $(1, n)$ and QGA_LSSVM separately and the predicted values are then regarded as the inputs of the combiner, which dynamically assigned non-linear weight coefficients to them. The simulation results indicate that the prediction accuracy of the combined prediction model is higher than those of single models and the running time is shortened. This new model is suitable for fault prediction with deterministic and random trends. In future research, the kurtosis predicted by the proposed model will be combined with the kurtosis threshold to achieve early warning of wind turbine misalignment faults.

Author Contributions: Y.X. and Z.H. contributed to paper writing and the whole revision process. All authors have read and agreed to the published version of the manuscript.

Funding: This research was funded by the National Natural Science Foundation of China (51577008).

Acknowledgments: The authors are grateful to the anonymous reviewers for their constructive comments on the previous version of the paper.

Conflicts of Interest: The authors declare no conflicts of interest.

References

1. Bandoc, G.; Shop, R.; Patriche, C.; Deprived, M. Spatial assessment of wind power potential at global scale. A geographical approach. *J. Clean. Prod.* **2018**, *200*, 1065–1086. [CrossRef]
2. Arshad, M.; O'Kelly, B. Global status of wind power generation: Theory, practice, and challenges. *Int. J. Green Energy* **2019**, *16*, 1073–1090. [CrossRef]
3. Koltsidopoulos, P.A.; Thies, P.R.; Dawood, T. Offshore wind turbine fault alarm prediction. *Wind Energy* **2019**, *22*, 1779–1788. [CrossRef]
4. Liu, Z.; Zhang, L. A review of failure modes, condition monitoring and fault diagnosis methods for large-scale wind turbine bearings. *Meas. J. Int. Meas. Confed.* **2020**, *149*, 107002. [CrossRef]
5. Xiao, Y.; Wang, Y.; Mu, H.; Kang, N. Research on Misalignment Fault Isolation of Wind Turbines Based on the Mixed-Domain Features. *Algorithms* **2017**, *10*, 67. [CrossRef]
6. Liao, M.; Liang, Y.; Wang, S.; Wang, Y. Misalignment in Drive Train of Wind Turbines. *J. Mech. Sci. Technol.* **2011**, *30*, 173–180.

7. Tonks, O.; Wang, Q. The detection of wind turbine shaft misalignment using temperature monitoring. *CIRP J. Manuf. Sci. Technol.* **2017**, *17*, 71–79. [CrossRef]

8. Kusiak, A.; Li, W. The prediction and diagnosis of wind turbine faults. *Renew. Energy* **2011**, *36*, 16–23. [CrossRef]

9. Liu, Y.; Qiao, N.; Zhao, C.; Zhuang, J. Vibration signal prediction of gearbox in high-speed train based on monitoring data. *IEEE Access* **2018**, *6*, 50709–50719. [CrossRef]

10. Kavousi-Fard, A.; Su, W. A Combined Prognostic Model Based on Machine Learning for Tidal Current Prediction. *IEEE Trans. Geosci. Remote Sens.* **2017**, *55*, 3108–3114. [CrossRef]

11. Liu, Z.; Zhang, L.; Carrasco, J. Vibration analysis for large-scale wind turbine blade bearing fault detection with an empirical wavelet thresholding method. *Renew. Energy* **2020**, *146*, 99–110. [CrossRef]

12. Zhao, H.; Zhang, X.; Guo, W. Early fault prediction of wind turbine drive train system. In Proceedings of the 6th IASTED Asian Conference on Power and Energy Systems, Phuket, Thailand, 10–12 April 2013.

13. Yang, Y.; Bai, Y.; Li, C.; Yang, Y. Application Research of ARIMA Model in Wind Turbine Gearbox Fault Trend Prediction. In Proceedings of the 2018 International Conference on Sensing, Diagnostics, Prognostics, and Control (SDPC), Xi'an, China, 15–17 August 2018.

14. Amarnath, M.; Shrinidhi, R.; Ramachandra, A.; Kandagal, S.B. Prediction of defects in antifriction bearings using vibration signal analysis. *J. Inst. Eng. Mech. Eng. Div.* **2004**, *85*, 88–92.

15. Wu, G.; Liang, P.; Long, X. Forecasting of Vibration Fault Series of Stream Turbine Rotor Based on ARMA. *J. South China Univ. Technol.* **2005**, *33*, 67–73.

16. Shi, M.; Jiang, L.; Fu, Y. Study on Prediction Methods for the Fault State of Rotating Machinery Based on Dynamic Grey Model and Metabolism Grey Model. *Wirel. Pers. Commun.* **2018**, *102*, 3615–3627. [CrossRef]

17. Yu, H.; Li, H.; Tian, Z.; Wang, Y. Rolling bearing fault trend prediction based on composite weighted KELM. *Int. J. Acoust. Vibr.* **2018**, *23*, 217–225. [CrossRef]

18. Zhang, K. *Research on Hub Bearing Fault Prediction Based on Combination Model*; Henan University of Science and Technology: Henan, China, 2017; Available online: http://cdmd.cnki.com.cn/Article/CDMD-10464-1017829413.htm (accessed on 10 January 2019).

19. Xu, Y. Rolling element bearing Fault prediction based on Grey Sequential Extreme learning machine. *Comput. Meas. Control* **2017**, *25*, 63–65.

20. Wang, P.; Li, Y.; Zong, X. The new forecasting method of combination based on SVM and grey. *J. Hebei Acad. Sci.* **2008**, *25*, 5–7.

21. Yan, H.; Lu, J.; Qin, Q.; Zhang, Y. A Nonlinear Combined Model for Wind Power Forecasting Based on Multi—attribute Decision—making and Support Vector Machine. *Autom. Elec. Power Syst.* **2013**, *37*, 29–34.

22. Zhang, A. *Research on Combination Forecasting Method of Regional Power Grid under Short-Term Load*; North China Electric Power University: Baoding, China, 2008; Available online: http://www.wanfangdata.com.cn/details/detail.do?_type=degree&id=Y1455060 (accessed on 10 January 2019).

23. Kong, X.; Liu, X.; Lee, K.Y. Data-driven modelling of a doubly fed induction generator wind turbine system based on neural networks. *IET Renew. Power Gener.* **2014**, *8*, 849–857. [CrossRef]

24. Wang, H.; Hu, D. Comparison of SVM and LS-SVM for regression. In Proceedings of the 2005 International Conference on Neural Networks and Brain Proceedings, Beijing, China, 13–15 October 2005.

25. Zhao, M.; Gao, H.; Xu, M.; Guo, Z.; Qiao, H.; Wu, X.; Huang, B. Application of multi-variable grey model for ball screw remaining life prediction. *Comput. Integr. Manuf. Syst.* **2011**, *17*, 846–851.

26. Huang, D.; Huang, L. Present Situation and Development Tendency of Grey System Theory in Fault Forecast Application. *J. Gun Launch Control* **2009**, *24*, 88–92.

27. Lou, Y.; Zhang, Q.; Liu, G.; Fan, Q. Application of grey multi-variable forecasting model in predicting urban water consumption. *Water Resour. Prot.* **2005**, *21*, 11–13.

28. Liu, S.; Xu, L.; Li, D.; Zeng, L. Dissolved oxygen prediction model of eriocheir sinensis culture based on least squares support vector regression optimized by ant colony algorithm. *Trans. Chin. Soc. Agric. Eng.* **2012**, *28*, 167–175.

29. Han, K.H.; Kim, J.H. Genetic quantum algorithm and its application to combinatorial optimization problem. In Proceedings of the 2000 Congress on Evolutionary Computation, La Jolla, CA, USA, 16–19 July 2000; IEEE Computer Society.

30. Zhang, X.; Sui, G.; Zheng, R.; Li, Z.; Yang, G. An Improved Quantum Genetic Algorithm of Quantum Revolving Gate. *Comput. Eng.* **2013**, *39*, 234–238.

31. Zhang, G.; Li, N.; Jin, W.; Hu, L. A Novel Quantum Genetic Algorithm and Its Application. *Acta Electron. Sin.* **2004**, *32*, 476–479. [CrossRef]
32. Jiang, S.; Zhou, Q.; Zhang, Y. Analysis on parameters in an improved quantum genetic algorithm. *Int. J. Digit. Content Technol. Appl.* **2012**, *6*, 176–184.
33. Sun, Y.; Ding, M.; Zhou, C.; Fu, Y.; Cai, C. Route planning based on quantum genetic algorithm for UAVs. *J. Astronaut.* **2010**, *31*, 648–654.
34. Bates, J.M.; Granger, C.W.J. The combination of forecasts. *Oper. Res. Q.* **1969**, *20*, 451–468. [CrossRef]
35. Shi, S. *The research of Demand Forecasting Model Based on Improved Grey System, BP Neural Network and Support Vector Machine*; Northeastern University: Shenyang, China, 2012. Available online: http://www.wanfangdata.com.cn/details/detail.do?_type=degree&id=J0119208 (accessed on 12 January 2019).
36. Li, X. *Research on Combined Forecasting Methods for Wind Power and its Application*; North China Electric Power University: Beijing, China, 2016. Available online: http://cdmd.cnki.com.cn/Article/CDMD-11412-1016267071.htm (accessed on 11 January 2019).
37. Li, G.; Liu, Z.; Li, J.; Fang, Y.; Shan, J.; Guo, S.; Wang, Z. Modeling of ash agglomerating fluidized bed gasifier using back propagation neural network based on particle swarm optimization. *Appl. Therm. Eng.* **2018**, *129*, 1518–1526. [CrossRef]
38. Vapnik, V.; Cortes, C. Support vector networks. *Mach. Learn.* **1995**, *20*, 273–297. [CrossRef]
39. Chanvillard, G.; Jones, P.J.; Aitcin, C.P. Evaluation of the statistical significance of a regression and selection of the best regression using the coefficient of determination R^2. *Cem. Concr. Aggreg.* **1993**, *15*, 31–38.
40. Xiao, Y.; Kang, N.; Hong, Y.; Zhang, G. Misalignment Fault Diagnosis of DFWT Based on IEMD Energy Entropy and PSO-SVM. *Entropy* **2017**, *19*, 6. [CrossRef]
41. Xiao, Y.; Hong, Y.; Chen, X.; Chen, W. The Application of Dual-Tree Complex Wavelet Transform (DTCWT) Energy Entropy in Misalignment Fault Diagnosis of Doubly-Fed Wind Turbine (DFWT). *Entropy* **2017**, *19*, 587. [CrossRef]
42. Li, C.; Guo, Y. Research on Performance Evaluation of Rolling Bearing Performance Based on SK and Other Indicators and SVM. *Electron. Sci. Technol.* **2020**, *33*, 6–12.
43. Zhang, L.; Rong, Y.; Liu, K.; Zhang, B. State pre-warning and optimization for rotating-machinery maintenance. *J. Univ. Sci. Technol. Beijing* **2017**, *39*, 1094–1100.
44. Guo, Q.; Wang, C.; Liu, P. Application of Time Domain Index and Kurtosis Analysis Method in the Fault Diagnosis of Rolling Bearing. *Mech. Transm.* **2016**, *40*, 172–175.
45. Cheng, X.; Wang, P. Fault Diagnosis of Rolling Bearing Based on Time Domain and Frequency Domain Analysis. *J. Hebei United Univ. Nat. Sci. Ed.* **2020**, *42*, 58–64.
46. Ma, Z.; Xiao, Q. The application of Mirror Extension for the end Effects in EMD Method. *Sci. Technol. Vis.* **2019**, *24*, 31–32.
47. Yu, Y.; Yu, D.; Cheng, J. A roller bearing fault diagnosis method based on EMD energy entropy and ANN. *J. Sound Vibr.* **2006**, *294*, 269–277. [CrossRef]
48. Zhang, C.; Chen, J.; Guo, X. Gear fault diagnosis method based on EEMD energy entropy and support vector machine. *J. Cent. South Univ.* **2012**, *3*, 932–939.
49. Jiang, J.; Wang, Q. Motor Bearing Fault Diagnosis Based on MEEMD and Kurtosis-Relevant Coefficient. *Electr. Drive* **2018**, *37*, 65–70.
50. Shi, H.; Hu, B. Survey of Dual-Tree Complex Wavelet Transform and Its Applications. *Inf. Electron. Eng.* **2007**, *5*, 229–234.
51. Ma, Z.; Wang, H.; Sun, Q. Sound Quality Prediction for Vehicle's Door-slamming Noise based on EEMD Sample Entropy and Wavelet Neural Network. *Noise Vibr. Control* **2019**, *39*, 122–127.
52. Tan, K.; Wang, W.; Zhang, L.; Liang, Y. Combination Forecasting of Coal Industry Human Resource Demand Based on Multiple Regression Model and Grey Forecasting. *Coal Eng.* **2019**, *51*, 151–154.
53. Huang, H.; Chen, Y.; Zhou, J.; Li, Z.; Yun, Y. Prediction and analysis of heat island intensity in mega city based on Grey System: A case study of Tianjin. *J. Arid Land Res. Environ.* **2019**, *33*, 126–133.
54. Zhou, Y. Land subsidence trend of Taiyuan City, Shanxi based on Grey Verhust Model. *Chin. J. Geol. Hazard Control* **2018**, *29*, 94–99.

Article

GA-Adaptive Template Matching for Offline Shape Motion Tracking Based on Edge Detection: IAS Estimation from the SURVISHNO 2019 Challenge Video for Machine Diagnostics Purposes

Alessandro Paolo Daga * and Luigi Garibaldi

Dipartimento di Ingegneria Meccanica e Aerospaziale—DIMEAS, Politecnico di Torino, Corso Duca degli Abruzzi, 24, I-10129 Torino, Italy; luigi.garibaldi@polito.it

* Correspondence: alessandro.daga@polito.it

Received: 11 December 2019; Accepted: 25 January 2020; Published: 29 January 2020

Abstract: The estimation of the Instantaneous Angular Speed (IAS) has in recent years attracted a growing interest in the diagnostics of rotating machines. Measurement of the IAS can be used as a source of information of the machine condition per se, or for performing angular resampling through Computed Order Tracking, a practice which is essential to highlight the machine spectral signature in case of non-stationary operational conditions. In these regards, the SURVISHNO 2019 international conference held at INSA Lyon on 8–10 July 2019 proposed a challenge about the estimation of the instantaneous non-stationary speed of a fan from a video taken by a smartphone, a pocket, low-cost device which can nowadays be found in everyone's pocket. This work originated by the author to produce an offline motion-tracking of the fan (actually, of the head of its locking-screw) and obtaining then a reliable estimate of the IAS. The here proposed algorithm is an update of the established Template Matching (TM) technique (i.e., in the Signal Processing community, a two-dimensional matched filter), which is here integrated into a Genetic Algorithm (GA) search. Using a template reconstructed from a simplified parametric mathematical model of the features of interest (i.e., the known geometry of the edges of the screw head), the GA can be used to adapt the template to match the search image, leading to a hybridization of template-based and feature-based approaches which allows to overcome the well-known issues of the traditional TM related to scaling and rotations of the search image with respect to the template. Furthermore, it is able to resolve the position of the center of the screw head at a resolution that goes beyond the limit of the pixel grid. By repeating the analysis frame after frame and focusing on the angular position of the screw head over time, the proposed algorithm can be used as an effective offline video-tachometer able to estimate the IAS from the video, avoiding the need for expensive high-resolution encoders or tachometers.

Keywords: machine vision; machine diagnostics; instantaneous angular speed; SURVISHNO 2019 challenge; video tachometer; motion tracking; edge detection; parametric template modeling; adaptive template matching; genetic algorithm

---

1. Introduction

Rotating machinery are fundamental components of mechanical systems for most of the industrial applications, as mechanical power is commonly obtained in the form of torque at a rotating speed. Electrical motors, internal combustion engines and motors in general, in fact, convert some sort of source energy into mechanical energy which is transferred to the final user through a mechanical transmission (e.g., a gearbox) which provides speed and torque conversion.

The speed is then a piece of essential system information, usually kept under control to accomplish a determined work. As a general consideration, feedback controllers can easily maintain their process variable (i.e., the speed in this case) close to its setpoint, but noise and disturbances may affect the controlled system instantaneous output causing the variable to depart from and oscillate around its desired value. The appearance of damages in the system has then repercussions on the instantaneous speed, which can be used as a source of information of the machine health condition per se.

The information about the instantaneous speed is fundamental for machine diagnostics also in other ways. In the field of Vibration Monitoring (i.e., a particularly successful kind of condition monitoring based on vibration records) for example, it is very common to use measured or estimated information about the Instantaneous Angular Speed (IAS) to perform the so-called Order Tracking [1]. In fact, most of rotating machines are affected by phenomena which are locked to particular angular positions (e.g., the intake of a 4 strokes diesel engine lasts from 0 to π radians of the main shaft and is periodic of 4π; a gear featuring a wheel with M teeth whose N^{th} tooth is damaged features an anomalous meshing pattern at angle $\frac{2\pi N}{M}$ of the wheel's shaft with a periodicity of 2π). Furthermore, in the case of non-stationary acquisitions, the spectral signature of the machine can only be found after synchronization of the Fourier analysis on the shaft rotation (i.e., the spectrum is not given as a function of the frequency, but as a function of the orders of a reference shaft). Unfortunately, commercial Data Acquisition Systems (DAQ) are not efficient in sampling at constant angular increments of the reference shaft, so that resampling algorithms are used, based on the additional information of the angular position of the reference shaft over time [2–5].

The growing interest of the signal processing community in the IAS information is demonstrated by the increasing number of publications involving the estimation of the instantaneous frequency as well as the creation of a special issue on IAS processing in the Journal of Mechanical Systems & Signal Processing [6] and of a conference dedicated solely to condition monitoring in non-stationary operations [7]. In addition, at the international conferences Surveillance 8 (in 2015) [8] and SURVISHNO (in 2019), contests about IAS estimation were organized. The IAS is in effect spreading in many fields of diagnostics of rotating machinery, especially for motors and gearboxes.

Particular applications for motors diagnostics involve electrical motors [9–11], internal combustion engines [12–16], and hydraulic engines too [17].

Gearbox diagnostics was also covered from a general point of view [18–21] as well as considering particular applications (e.g., windmills [22]) or focusing on relevant components such as the bearings [23,24].

Machining processes can also be subject to diagnostic analysis through the IAS estimation. The milling cutting force, for example, is proved to be reflected by the IAS [25,26].

1.1. Instantaneous Angular Speed (IAS) Review

The IAS information is, in simple terms, a measure of the rotation speed of a rotating component of a machine defined at an angular resolution corresponding to at least one value per revolution [26]. From a physical point of view then, the IAS is defined starting from the angular position $\alpha(t)$ of a shaft as:

$$\omega(t) = \dot{\alpha}(t) = \frac{d\alpha}{dt} \tag{1}$$

A measure of this quantity can be obtained using analog sensors such as the tachometer-dynamo which translates the rotational speed into an electrical signal, or with analog angular position sensors such as the resolver. Nevertheless, such devices can nowadays be considered obsolete and are often substituted by more reliable digital sensors, whose signals can be processed to produce an estimate:

$$\hat{\omega}(t) \cong \frac{\Delta\alpha}{\Delta t} \; ; \; \hat{\omega} \to \omega \; for \; \Delta t \to 0 \tag{2}$$

In particular, two strategies can be put in place. One involves the measurement of the angle $\Delta\alpha$ swept in a constant interval of time Δt by the shaft. Nevertheless, the angular measurements can be affected by larger errors than the time measurements' ones. A second more widespread and accurate method consists of measuring the time elapsed between successive pulses (Elapsed Time $ET = \Delta t$) corresponding to a known swept angle $\Delta\alpha$ which is a characteristic of the sensor (e.g., an encoder). An analysis of the resolution and the speed estimation error can be found in [27].

In any case, all these methods involve the use of an additional sensor (generally referred to as a generic "tachometer" but not to be confounded with the tachometer-dynamo sensor). The interest of the scientific diagnostics community then has recently moved to IAS estimation from the easily available accelerometric signals, fundamental for vibration monitoring, going "encoder-less" or "tacholess". In this regard, several different approaches are possible [7].

In particular, the simplest idea is to track a shaft-speed related harmonic (possibly showing a good Signal to Noise Ratio-SNR) from its corresponding peak in a time-frequency representation of the signal (i.e., a Spectrogram, computed for example via a Short Time Fourier Transform). This can be made more accurate by averaging the tracks from multiple harmonic orders, which can also be automatically selected by the algorithm (e.g., Multi-Order Probabilistic Approach-MOPA, Ceptrsum-based MOPA, or ViBES).

A second approach to the IAS estimation is the demodulation of a shaft-speed related harmonic exhibiting good SNR. This follows the idea that vibration signals can be modulated by the revolution of the shaft so that phase demodulation can recover the shaft speed from the band-pass isolated harmonic of interest.

Demodulation can also be performed using the Teager-Kaiser Energy Operator.

Finally, the two main procedures of tracking and demodulation can be exploited together as implemented in the Vold-Kalman filtering.

At any rate, both the measurement and estimation of the IAS have some limitations. In the first case, the use of high-resolution encoders allows to get reliable and accurate estimates of the IAS, but the angular sensors can be very expensive and need to be mounted on a shaft added for the purpose. On the other hand, the second case exploits the cheap and reliable accelerometers which can be added on a machine without special design updates. The estimated IAS, however, is less accurate and less reliable, and needs computational time.

In this article then, the measurement of the IAS is tackled at the scope of testing an offline, low-cost video-tachometer approach, as suggested by the SURVISHNO 2019 challenge (see Supplementary Material). The question was "how far is it possible to carry out relevant analysis from a video or a rotating fan acquired by a smartphone which can reach a maximum rate of 30 frames per second?".

In order to answer the question, the field of computer vision was explored so as to point out the main approaches to shape detection and image recognition at the scope of extracting the angular position $\alpha(t)$ of the fan from the video and then obtain an estimate of the IAS.

1.2. Brief Literature Review of Computer Vision

Computer (or Machine) Vision is a scientific discipline that deals with Machine Learning applied to digital images and videos coming from cameras as well as any other visual representation derived from various sensors such as ultrasonic cameras, range sensors, radars, tomography devices, etc. The objective is transforming visual images into descriptions of the world by extracting data and features, producing then information which is fundamental in decisional processes. Applications of Machine Vision to industrial processes include automatic inspection (e.g., to detect manufacturing defects), surveillance and security (e.g., detection of events, face recognition, etc.), motion control, navigation and human-machine interaction (e.g., robots, autonomous vehicles, etc.), modeling objects or environments, improving human vision (e.g., medical image analysis and detection), or organizing information (e.g., image classification databases, etc.).

Anyway, according to the need of the particular application, this article is interested only in a small sub-domain of Machine Vision algorithms, namely object recognition (i.e., finding and identifying objects in an image or video), video tracking (i.e., locating a moving object over time using a camera), and motion estimation (i.e., determining motion of a body from two following 2D frames).

In order to obtain the angular position of the fan $\alpha(t)$, in fact, all the three domains are needed. First, the subject must be identified in a frame. Then, its position and orientation should be compared to the subject in previous or in a reference frame. This way the motion can be tracked from frame to frame. In general, two main categories of algorithms can be found in the literature [28–30]. Direct methods are based on the pixel information, while indirect methods make use of features such as corners or particular points of the subject. Feature-based methods minimize an error measure based on distances in the feature space, while direct methods minimize an error measure based on direct image information collected from all pixels in the image (i.e., typically the pixel brightness, which is usually computed from the RGB color image at a pre-processing stage).

Complete reviews of approaches to image processing and face recognition (i.e., a common application of object recognition) can be found in [31–33] from which it is clear that Neural Networks strongly entered the game of machine vision (in particular for feature-based methods). However, all the reviews also agree in pointing out a common foundation of the processing: Template Matching (TM).

Template Matching [34] is a technique in digital image processing for finding small parts of an image which match a template image (i.e., an image designed to serve as a model). TM is often considered a very basic and limited approach to computer vision, but it is actually involved in many old and new techniques. In [35] for example, TM is taken into account only for reference-comparison inspection of mass-produced integrated circuits precisely aligned on a conveyor (i.e., equal objects, with the same location, scale, and orientation in the image). In fact, the limitations of traditional TM are well known [34–36] and can be summarized as: (a) noise, illumination changes, and occlusions, (b) background changes and clutter, (c) rigid and non-rigid transformations and scale changes (i.e., images are a projection of a 3D scene onto a 2D plane), (d) high computational cost.

Point (a) is commonly tackled first at a pre-processing stage, when the RGB color image is translated into a pixel-wise single channel brightness information (i.e., the luminance) and edge detection is performed, and later by focusing on the stability and robustness of the selected similarity measure, which affects also point (c). Examples of standard similarity measures are the pixel-wise sum of differences of the search image and the template, the sum of products (or cross-correlation) among them, the Best-Buddies-Similarity based on Nearest-Neighbor matches of features, the Deformable Diversity Similarity [36]. Limitations in (b) and (d) are often faced with the simple trick of using masks to remove non-interesting areas (i.e., providing a search window), or otherwise, reducing the number of sampling points by decreasing the resolution. Point (c) was sometimes dealt with the implementation of multiple templates with different scales and rotations (e.g., eigenspaces) or with Deformable Part Models (DPM) or Deformable Template Matching [37]. This paper anyway focuses on the exploitation of the Genetic Algorithm for dealing with rigid transformations and scale variations in the template.

1.3. GA and Template Matching: A Review

Evolutionary Algorithms are recently enjoying new success in the scientific community for generating good solutions to optimization and search problems without relying on assumptions about the underlying fitness landscape (i.e., they perform derivative-free optimization).

A Genetic Algorithm (GA) is a heuristic algorithm inspired by Charles Darwin's theory of natural evolution via natural selection, where the best individuals are more prone to reproduction and have better offspring.

When focusing on TM, GA can be found in several applications and in different fields, from manufacturing (i.e., Integrated Circuits quality inspections) to security and surveillance (i.e., animals recognition or face recognition) up to medicine (i.e., nodulus recognition in Computed Tomography CT scans) as found in [38–42]. In particular, in [38] GA is used to speed up the TM by shrinking the image to be processed. More refined employment of GA can be found in [39], where face recognition is performed by TM using a T-shaped template isolating eyes, nose and mouth, resized by GA to find a better match when the size of search image and template is different. In [40] further improvement is proposed, as a Deformable Template generated as a Point Distribution Model (PDM) is adapted by GA to measure characteristic landmark points (i.e., vertices, nodes, markers, etc.) on cattle images for morphological assessment of bovine livestock. The idea of generating a template through a model optimized by GA was applied also in [41] and in [42] for automatic detection of lung nodules in chest spiral CT scans. Nevertheless, all three applications show some weaknesses. In [40], because of the scope of the analysis, the deformable template is GA adapted to locate landmark points on the cattle PDM profile, but this does not allow to extract clear and unique information about scale, orientation, or center of the template shape, which remains a parametric function (i.e., the template is not a digital image). In [41], a two-step GA is proposed for real-time shape tracking. The first step optimizes the template which is not directly generated by a mathematical model but comes from a mathematical mask (characterized by 3 parameters) on a digital image template, while the second optimizes orientation and center position the template (but does not accounts for the scale). Nevertheless, the higher computational efficiency of using a two-steps GA rather than a single step GA is not justified. Finally, in [42], the template is actually halfway between the parametric function and the digital image: a simple parametric function is used in this case to generate multiple images which will be used as multiple templates (different scale and rotation). Nevertheless, GA is set up to perform a discrete rather than a continuous search of the best matching template in terms of rotation, scale, and center position. This limits the potential of GA-TM of getting sub-pixel accuracy [30].

As a result of these considerations, a novel GA-adaptive TM technique is proposed in this work. In particular, GA is integrated into the TM so as to reconstruct a digital template from a simplified parametric geometrical model (which acts as a mask) whose parameters can be continuously optimized in terms of scale, rotation, and center position so as to maximize a similarity measure and to find the best match. This not only overcomes the well-known issues of the traditional TM related to deformations in the search image with respect to the template but enables to resolve the position of the center at a resolution that goes beyond the limit of the pixel grid, allowing an effective shape tracking which is used in this paper for implementing an offline video-tachometer able to estimate the IAS of the SURIVSHNO 2019 fan from the video, avoiding the need of expensive high-resolution encoders or tachometers.

2. Materials and Methods

This work is meant to propose an inexpensive but effective video-tachometer using a 30 fps (i.e., frames per second) video from a mobile phone. The target is the IAS estimation of the SURIVSHNO 2019 fan. The raw dataset is then composed by a bunch of sequential digital color pictures, each of them corresponding to a matrix of pixels (i.e., "picture element": the smallest addressable element of the digital image) updated during a full scan of the camera image sensor. The dataset is described in Section 2.1. In Section 2.2 the principle of TM is introduced, while in Section 2.3 the GA is integrated into TM. Finally, the overall methodology for estimating the IAS is reported in Section 2.4.

2.1. Data Description: the SURVISHNO 2019 Challenge Video and Its Critical Issues

As already introduced, Computer Vision deals with the understanding and interpreting of visual representations such as digital images and videos, as well as other representations which will not be considered in this paper. A video is the electronic medium for recording and displaying a moving visual media, namely a chronographic sequence of photographic shots which forms a representation of the visual world and can capture motion. The representation of the visual characteristics of an object is converted by image sensors into digital signals that can be processed by a computer and made output through a screen as a visible-light image. The 2D digital image is spatially discretized in a number of addressable elements (i.e., the pixels-px) organized in rows and columns to cover the entire image space. The standard full-HD High-Definition Television (HDTV) system uses a resolution of 1920×1080 px with 16:9 aspect ratio, so that each of the 2,073,600 pixels stores three-channels color information. Trichromacy, in fact, mimics the animal vision which uses three different types of cone cells in the eye to perceive not only light intensity but also its spectral composition (i.e., color). A very common set of primary colors is that defined by the RGB color model, an additive model in which red, green, and blue light are added together to reproduce a broad range of colors. In particular, graphics file formats usually store RGB pictures as 24-bit images (i.e., RGB24 format), where RGB components are 8 bits each, so that each color intensity can be rendered at 256 levels (normalized to unity between 0 and 1 or, more commonly, with integers between 0 and 255), leading to a potential of 16,777,216 (about 16 million) colors.

Therefore, in digital imaging systems (e.g., digital cameras, mobile phones, etc.) the acquisition corresponds to an interrogation of each pixel photo-sensor so that the full image is recorded; to produce a video this image recording is repeated in time at a given sampling frequency (commonly 30 fps). Nevertheless, the pixel sensors can be either interrogated simultaneously (i.e., global shutter) or, more frequently in mobile phones, one after the other in a predetermined sequence (i.e., rolling shutter).

In the particular case, the SURIVSHNO 2019 video is a sequence of 1298 RGB24 images acquired at 30fps for a duration of 43.3 s. All the frames are recorded at a full-HD resolution of 1920 × 1080 px with 16:9 aspect ratio. The RGB color image depicts a front view of a fan composed by 10 blades, coupled to a spindle by a hexagonal shaft and an 8.8 screw. The first frame of the video is reported in Figure 1.

Figure 1. The first frame of the video, digitalized as a 1080 × 1920 × 3 matrix: full-HD resolution of 1920 × 1080 (16:9 aspect ratio), where each pixel color is represented as a triplet of intensities for the three-color components namely red, green, and blue (RGB additive trichromacy color model).

The IAS of the fan is unknown, nevertheless, it can be recognized as strongly non-stationary by watching the video. Some typical issues of non-stationary cases, in fact, arise. In particular, spatial aliasing due to the rolling shutter effect of the camera can be easily noticed. As the fan is spinning counterclockwise at an increasing rotational speed, the blades on the left side appear to get thicker while the blades on the right side appear to become thinner as the video progresses in time. This is visualized in Figure 2, in contrast to Figure 1.

Figure 2. The thousandth frame of the video. The rolling shutter effect is highlighted as a deformation of the blades.

Aliasing is a typical issue of digital signals. A digitally reconstructed image, in fact, will differ from the original image (i.e., analog) because of the spatial discretization (i.e., the sampling) so that visible patterns or deformations can compromise the quality of the reconstruction.

Temporal aliasing, determined by the sampling frequency or, in case of videos, by the frame rate of the camera, is a major concern of Digital Signal Processing. In videos, because of the limited frame rate (N.B., limited with respect to the rotating speed of the object), a rotating object like a fan or a wheel looks like turning in reverse or too slowly. A similar effect is probably experienced by any human beings in the form of an optical illusion called "wagon-wheel effect" which may occur even under truly continuous illumination because of the human visual perception.

Sampling at 30 fps, in accordance with the Nyquist sampling theorem, allows to correctly picture phenomena which are bandlimited to half the sampling rate (i.e., 15 Hz, the Nyquist frequency) without aliasing. In the SURIVSHNO 2019 video, the fan starts from a standstill and accelerates up to values lower than the Nyquist frequency, so that temporal aliasing does not occur. Nevertheless, by looking at the video, the optical illusion of a reverting direction of rotation occurs anyway as the brain cannot recognize the 10 equal blades of the fan, so the exceeding of 1,5 Hz causes a reversal of the perceived direction of rotation.

A final issue is related to the autofocus of the camera. When taking photos, in fact, a convex lens is used in the camera to focus incoming light onto a photo-sensor array (e.g., a Complementary metal-oxide-semiconductor-CMOS-photo-sensor). In order to ensure crisp and clear images, the optical system commonly uses a control system and a motor to optimize the distance between the lens and the sensor. This can obviously lead to distortions of the image during the video.

To summarize, three main issues should be tackled:

- Spatial Aliasing related to the rolling shutter effect,
- Temporal Aliasing due to the 30-fps sampling rate given the 10 equal blades of the fan,
- Additional autofocus distortions.

Nevertheless, a workaround can be found to simplify things. First, the spatial aliasing occurs when the object moves faster than a limit speed dictated by the rolling shutter clock. Being the fan is rotating around its center, the higher the distance from the center, the higher the tangential speed, so that, focusing on a part of the image very near to the center (i.e., the fan-locking screw head), the spatial aliasing effect is minimal.

Second, the temporal aliasing, in this case, is more a visualization issue rather than a real problem for the analysis. The fan speed, in fact, is always lower than 10 Hz, so that if the attention is brought to a feature that occurs just once per revolution (i.e., the 8.8 logo on the fan-locking screw head) rather than the blades (which are 10 and not distinguishable), no temporal aliasing occurs.

Finally, the autofocus distortions can be accounted for, together with other perspective distortions by implementing the adaptive TM introduced in the following sections. In the SURVISHNO 2019 challenge video acquisition, in fact, the camera was almost but not perfectly aligned to the fan axis, so that, during the revolution, the center of rotation moves in a small region, while the image undergoes slight deformations.

2.2. Matched Filters and Template Matching

The problem of finding parts of a search image which match a template image is just a bi-dimensional extension of the common unidimensional Signal Processing (SP) problem of detecting the presence of a template signal in a search signal, typically a noise-affected measurement. This problem was first solved in the mid-40s by North, Vleck, and Middleton [43–46] as a response to the immediate need to improve radar performance during World War II [43]. In the original framework, the issue with radar is to highlight the presence of an echo (i.e., a known template) exhibiting little power and obscured by noise in a received signal (i.e., the search signal). Assuming a Gaussian white noise (i.e., with a flat power spectrum), the noise contributes with equal undesired power at all frequencies, while the signal, on the contrary, shows a bandlimited spectral content. Considering a transmitted pulse (i.e., a rectangular function), its spectrum is described by a sinc pulse (i.e., a sinc function) which is theoretically defined over the whole frequency axis but has practically most of the power bound to low frequencies. A matched filter is then a linear time invariant filter that maximizes the signal-to-noise power ratio highlighting then the presence of the template in the search signal. Intuitively, it is then a filter that emphasizes the frequency where the template power is contained (i.e., the low frequencies) while attenuating those where the only noise is present. If the template is known then, it is sufficient to use the template spectrum for designing the best filter frequency response.

The filter impulse response to be convolved with the search signal (i.e., for discrete signals, $s[n]$, where n is the sample index related to time by the sampling period) is then just the time-reversed version of the template $\hat{t}[n]$ of finite length N (N.B., more in general, for complex search signals, the conjugated time-reversed). The filter is then said to be "matched" to the template. See Equations (3) and (4) for the discrete Matched Filter impulse response definition ($h[n]$).

In the Matched Filter output a peak occurs (i.e., the amplitude goes "considerably" greater than the rest of the output signal $y[n]$ in the time domain) when the template signal is detected. By playing a bit with the notation, the convolution of the search signal with the time-reversed version of the template ($s[n] * h[n]$ in Equation (5)) is equivalent to the cross-correlation of the template (as it is) with the search signal ($r_{ts}[n]$ in Equation (6)). In the same way, if the cross-correlation shows a peak for a given delay (or lag) k, then the template is detected.

$$
\begin{cases}
 & s[k] \quad -K < k < K \\
t[n] = \hat{t}[n] & n = 0,\ldots,N-1 < K \\
t[n] = 0 & -K < n < 0 \vee N-1 < n < K
\end{cases}
\tag{3}
$$

$$
h[n] = t[-n]
\tag{4}
$$

$$
y[n] = s[n] * h[n] = \sum_{k=-K}^{K} h[n-k]s[k]
\tag{5}
$$

$$
y[n] = \sum_{k=-K}^{K} t[k-n]s[k] = \sum_{k=-K}^{K} t[k]s[k+n] = r_{ts}[n]
\tag{6}
$$

The same consideration holds also when the problem is extended to a discrete 2D search signal such as a monochromic image where the light intensity $s[n,m]$ is a function of the spatial coordinates n

and m over the pixel grid (i.e., a matrix). This problem is referred to as TM. Given a known template $\hat{t}[n, m]$ of size $N \times M$, it is possible to perform cross-correlation by simply moving the center of the template over each pixel of the search image pixel-grid of size $J \times K$ and calculate the sum of the pixel-wise products of s and $\hat{t}$ over the area spanned by the template $\hat{t}$. In a more rigorous formulation:

1. The template $\hat{t}$ is placed at $(n_0 + N/2, m_0 + M/2)$ in a matrix t of the same size $[n, m]$ of the search matrix $s[n, m]$ (Equation (7)),
2. The entrywise product (also known as Hadamard or Schur product, here "$\circ$") is performed finding the matrix $st[n, m]$ (Equation (8))
3. The correlation r_{ts} at $(n_0 + N/2, m_0 + M/2)$ is obtained by summing all the components in $st[n, m]$ (Equation (9))
4. By letting n_0 and m_0 vary in the range $n_0 = 1, \ldots, J - N + 1 \; \vee \; m_0 = 1, \ldots, J - M + 1$, the whole cross-correlation matrix is computed.

$$
\begin{cases}
t[n, m] = \hat{t}[n, m] \\
t[n, m] = 0
\end{cases}
\begin{aligned}
&s[j, k] \quad 1 < j < J \vee 1 < k < K \\
&n = n_0, \ldots, n_0 + N < J \,; \; m = m_0, \ldots, m_0 + M < K \\
&1 < n < n_0 \vee n_0 + N < n < J \,; \; 1 < m < m_0 \vee m_0 + M < m < K \\
&n_0 = 1, \ldots, J - N + 1 \vee m_0 = 1, \ldots, J - M + 1
\end{aligned}
\tag{7}
$$

$$
st[n, m] = s[n, m] \circ t[n, m] \tag{8}
$$

$$
r_{ts}[n_0 + N/2, m_0 + M/2] = \sum_{j=1}^{J} \sum_{k=1}^{K} st[j, k] \tag{9}
$$

If the template is present in the search image, the cross-correlation features a maximum. The template detection becomes then a search for the maximum. This implies that the cross-correlation can be used as an effective similarity measure.

Nevertheless, this basic approach to TM is effective only when the template is a crop from an acquired reference image, and the search image is acquired under the same conditions (i.e., illumination, scale, orientation, fixed background, etc.). That is why in [35] TM is considered only for quality inspection of precisely aligned integrated circuits.

The limitations of traditional TM, as already introduced in Section 1, are well known [34–36] and are here summarized:

(a) noise, illumination changes, and occlusions in the search image,
(b) background changes and clutter,
(c) rigid and non-rigid transformations, rotations, and scale changes (i.e., images are a projection of a 3D scene onto a 2D plane),
(d) high computational cost.

Nevertheless, in this particular application (i.e., the SURVISHNO 2019 fan), as introduced in the previous sub-section, it was proposed to solve the issue of aliasing by focusing on the fan-locking screw head region. This reduction of the search space performed by cropping the image around the center of rotation is beneficial also according to points (b) and (d), as the background is effectively removed, while the computational burden is lightened.

The overall pre-processing is described in the next subsection, which treats about the edge detection performed on the basis of the monochromic image obtained from the brightness of the original RGB24 image. This helps in relieving the issues in point (a).

Finally, point (c) is addressed by exploiting the GA for dealing with rigid transformations and scale variations of the template to obtain a better match.

2.3. Image Preprocessing

In order to maximize the performance of the TM, a preprocessing of the image is essential. Three fundamental steps were chosen, based on the literature [47]:

- Image cropping
- Gray monochrome conversion and image binarization (thresholding)
- Edge Detection

The first step is fundamental for removing the issues related to the background and in particular for improving the computational speed.

The second step is used to prepare the image for edge detection, limiting the effect of noise, illumination changes and occlusions in the search image. In this analysis, in fact, the edges are used as features to enhance the TM. The hybridization of template-based and feature-based approaches is not new (e.g., [48,49]), and allows to overcome the well-known issues of the traditional TM related to scaling and rotations of the search image with respect to the template. This will be the subject of Section 2.4.

2.3.1. Image Cropping

The image cropping is meant to remove the background, solving issues related to changes and clutter in part of the image of little relevance. Furthermore, decreasing the overall number of pixels in the image, the computational burden is reduced.

In this analysis, the image is shrunk from a matrix of 1080×1920 px to a matrix of 191×191 px, by cropping in the square region defined by the row indices in the range $475 \div 665$ px and the column indices in the range $485 \div 675$ px. The result is presented in Figure 3 for two frames of the video.

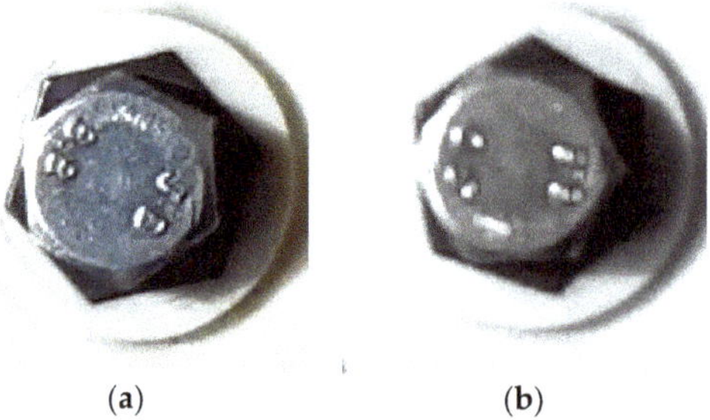

(**a**)   (**b**)

Figure 3. First (**a**) and thousandth (**b**) frames of the video. Notice how the image (**b**) is out of focus and the center of the screw is translated upwards. N.B., The selected x coordinate goes from left to right, while the y coordinate goes from the top to the bottom, differently from the standard.

2.3.2. Gray Monochrome Conversion and Image Binarization (Thresholding)

Conversion of an arbitrary color image to grayscale is not a unique procedure in general, as a different weighting of the color channels can effectively represent the effect of shooting black-and-white film. A common strategy is to use the principles of photometry and colorimetry to calculate the grayscale values so as to have the same relative luminance (i.e., the density of luminous intensity per unit area in a given direction) as the original color image. Given the RGB intensities (i.e., values in the range $0 \div 255$ for the RGB24 file format or normalized to $0 \div 1$) provided by the three channels R, G, and B, the luminance Y is defined as a weighted sum of these components. In this analysis, the coefficients from the ITU-R Recommendation BT.601 standard, revision 7 [50] are taken, so as to find:

$$Y = 0,299\ R + 0,587\ G + 0,114\ B \tag{10}$$

The formula reflects the eye color photoreceptors sensitivity, which has a maximum in the green-light region. Notice that, in general, human-perceived luminance is commonly referred to as brightness, while luma is the luminance of an image as displayed by a monitor.

The so obtained grayscale image is displayed in Figure 4a.

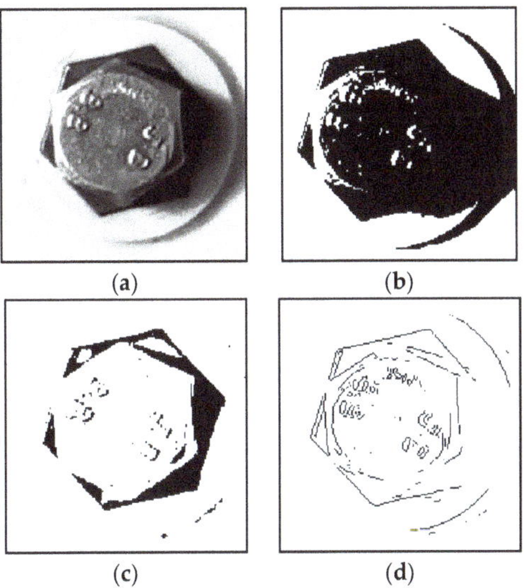

(a)   (b)
(c)   (d)

Figure 4. (**a**) Original grayscale: Grayscale conversion of the first frame of the video (Figure 3a) using luminance [50]; (**b**) Otsu threshold: Comparison of Otsu's thresholding [51]; (**c**) Adaptive thresholding (Bradley) [52,53] with sensitivity: 0,99; (**d**) Not-Sobel filtering [54,55] applied to the grayscale image in (a).

Once the gray monochrome image is obtained, thresholding is implemented. The goal of thresholding is to classify pixels as either dark (0) or light (1) to produce a black and white (i.e., binary) image based on the luminance information.

In its simplest implementation, the threshold is a constant set by the user, and the pixels' luminance is compared against this value. An automatic selection of the threshold was implemented by Otsu [51] as a Fisher's Discriminant Analysis performed on the intensity histogram. Otsu's threshold is then determined by minimizing intra-class intensity variance, or equivalently, by maximizing inter-class variance (N.B., the two classes are obviously dark vs. light).

Nevertheless, illumination changes in the image can lead to a bad classification. In this case, an adaptive threshold such as the Bradley's could perform much better [52]. The idea is to use a local threshold which can vary within the image as it is adapted to the average of surrounding pixels. Typically, a moving window of approximately 1/8th of the size of the image is used for computing the local mean intensity. Matlab implementation [52] also allows to tune the threshold using a scalar "sensitivity" in the range $0 \div 1$: high sensitivity value leads the thresholding of more pixels as foreground (i.e., class 1, light), at the risk of including some background pixels (i.e., class 0, dark).

Thresholding, in fact, is commonly used to separate foreground objects from their background, reinforcing the action of the image cropping.

Comparing Figure 4c to Figure 4b, the robustness of Bradley's adaptive thresholding to illumination changes is highlighted. Furthermore, it can be noticed that the circle in the image background is removed, improving the robustness to background changes and clutter.

2.3.3. Edge Detection

The objective of edge detection is to find the locations in a grayscale image where the change in intensity (i.e., Y) is sufficiently large to be taken as a reliable indication of an edge [35]. One of the most common detectors is the Differential Gradient edge detector called Sobel–Feldman filter [54], which uses two 3×3 windows convolved with the image to produce two directional pieces of information (i.e., approximated gradients) added to find the resulting magnitude. Finally, the magnitude information undergoes thresholding to produce a binary image of the edges (automatic heuristic threshold selection [55]), as reported in Figure 4d, where the logical not operator is applied to highlight the edges in black. As can be easily noticed in the picture, the edges are filtered and isolated very effectively, but the illumination affects the result.

By comparing Adaptive thresholding and Sobel filtering (i.e., Figure 4c vs. Figure 4d) it is clear how robustness to illumination changes is important in the analysis. Hence, in this work, the edge detection is left to Bradley's adaptive thresholding, to produce a search image with thicker edges more robust to noise, illumination changes, and background clutter (i.e., Figure 4c).

To summarize, the finally selected preprocessing is reported in Figure 5.

Figure 5. Preprocessing scheme.

2.4. GA-adaptive Template Matching

In traditional TM, as described in Section 2.2, the template is selected as a cutout from one larger reference search image and compared to all the successive test search images using the cross-correlation. The position of maximum correlation testifies the match, proving the template detection. This obviously works very well in case of a fixed framing camera depicting an object which translates on a plane orthogonal to the optical axis of the camera. Nevertheless, in case of rotations, scale changes or non-rigid transformations (N.B., images are a projection of a 3D scene onto a 2D plane, so that movement on a plane non-orthogonal to the optical axis of the camera can lead to deformations), the method cannot be used unless some technical device is implemented, such as using multiple templates with different scales and rotations (e.g., eigenspaces) or using Deformable Part Models (DPM) or implementing Deformable Template Matching [37]. Nevertheless, in this particular work, the Genetic Algorithm was selected to deal with rigid transformations of the template. The GA was used for adapting a parametric template so as to get the maximum correlation (i.e., the best match). The complete cross-correlation function (i.e., correlation for all the delays or lags) is never computed in this case; the GA is exploited for optimizing at the same time not only the scale and the orientation but also the location of the parametric template in the search space, so as to obtain an hybrid of the template-based and feature-based approaches which allows to overcome the issues of the traditional TM (i.e., scaling and rotations).

Notice that, in the SURVISHNO video, the framing is fixed, but the fan revolution occurs in a plane non-perfectly orthogonal to the optical axis, so that, during the revolution, the center of rotation of the fan moves in a small region, while the image undergoes slight deformations. These perspective-related deformations are neglected by the here-introduced algorithm, but relative translations of the locking screw head hexagon and the underlying hexagon (lying on two different planes) are allowed by breaking the template adaptation in three successive GA steps.

2.4.1. Template Parametric Model

The template parametric model arose from the exploitation of the geometrical features of the search image. In particular, three characteristic features were defined in order to determine the angle of rotation of the fan. The first two are related to the regular hexagonal shape of the screw head and the underlying driving shaft. The third is the resistance class logo (i.e., 8.8), which enables to discern the orientation of the screw and consequently that of the fan.

The three characteristic features are highlighted in Figure 6.

Figure 6. Geometrical edge features highlighted on the first frame of the video (Figure 3a): Outer hexagon model in red and its center in green, inner hexagon model in magenta and its center in cyan, 8.8 logo model in yellow, and the finally tracked vertex in blue.

Two parametric models for the edges are then built. The first represents a hexagon inscribed in a circumference and is governed by the coordinates of the center (i.e., the location), the radius of the circumference in which the hexagon is inscribed (i.e., the scale), and the angle of rotation of the hexagon. The second is the 8.8 logo, modeled as 5 circles, around one of the diagonals of the screw hexagon, as reported in yellow in Figure 6. This is governed by five parameters: a size parameter ruling the radii of the circles and the height of the writing, a width parameter giving the distance of the two 8 characters, a shift parameter allowing uneven positioning of the two 8 characters around the main axis, a radial distance of the writing from the center of the screw hexagon (either positive or negative to cover both sides of the reference axis with respect to the center of the screw), and an angular deviation from the reference axis, whose information is considered as a known input given the desired diagonal of the screw.

Given these two geometric models, a binary template of size 191×191 px can be produced as a function of these 13 parameters plus thickness information. The characteristic parameters are reported in Table 1. The ideal path from the model, defined in the continuous pixel space, can be used as a mask for lighting (i.e., turning to 1) the pixels covered by such a filter. The path thickness is obviously a relevant parameter, but to avoid overcomplicating the model, the thickness was pre-set to a constant value of 30 px for the hexagons (t and t' in Figure 7), while it is related to the scale parameter s for the 8.8 logo (*thickness* $= r_1 - r_2 = 3,5\,s$).

Table 1. Characteristic variables of the three parametric templates. The 13 independent variables are highlighted in red, *t* and *t'* parameters are constants, while the other parameters are derived.

Parameter	Description	Parameter	Description
X_c, Y_c	Center of the outer hexagon (OH)	R''	Distance of 8.8 logo from (X'_c, Y'_c)
R	Radius of the inscribing circle (OH)	$d\theta$	Deviation from *ax* slope direction
θ	Rotation of the OH	$s = r_2$	Logo size = hollow circles radii
t	Thickness of OH	r_1	Logo's circles radii $r_1 = 4,5s$
X'_c, Y'_c	Center of the inner hexagon (IH)	r_3	Logo's dot radius $r_3 = 2,25s$
R'	Radius of the inscribing circle (IH)	h	Logo's height $h = 5,5s$
θ'	Rotation of the IH	w	Logo's width $w = w_1 + w_2$
t'	Thickness of the IH	w_r	Logo's width ratio $w_r = w_1/w$
ax	8.8 intercepting diagonal of IH	w_1, w_2	Distance of "8" from Logo's dot

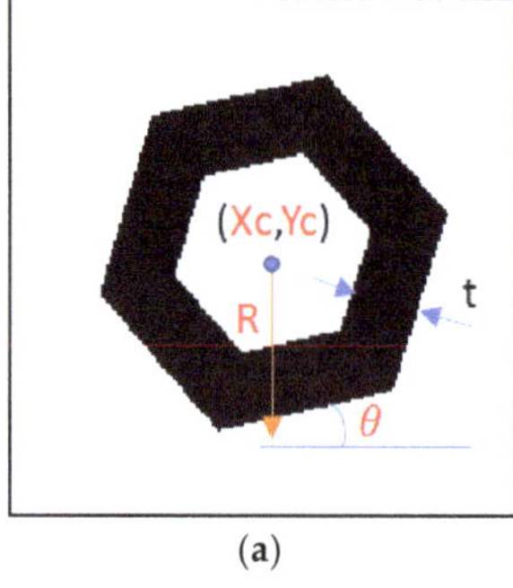

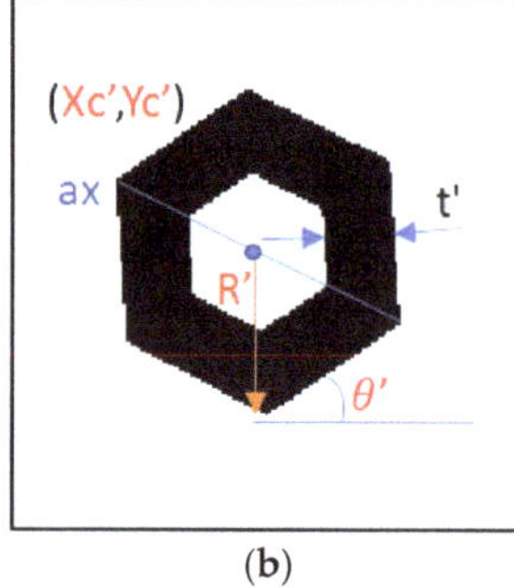

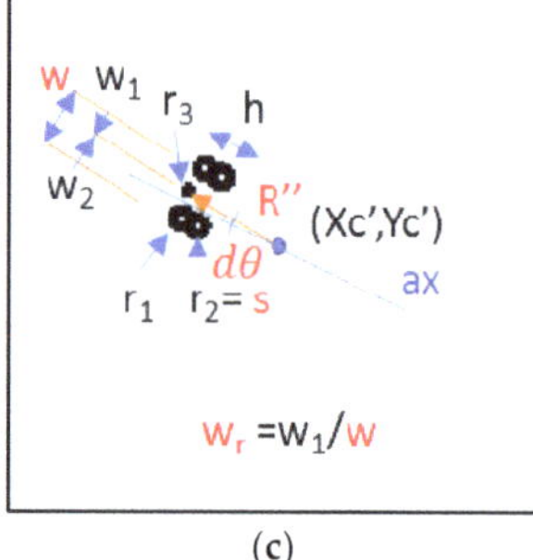

(a)   (b)   (c)

Figure 7. Binary templates after Genetic Algorithm (GA) optimization on the first frame (Figure 3a) with characteristic parameters highlighted in red (overall, 13 independent parameters, as reported in Table 1): (**a**) Outer hexagon template; (**b**) Inner hexagon template; (**c**) 8.8 logo template. N.B., The pictures "quantization" effect is determined by the 191 × 191 px grid of the search image, which dictates the final template resolution.

As highlighted in Figure 7, three different templates were actually generated, as the template adaptation was performed in three different subsequent GA optimizations, exploiting in the following steps the knowledge acquired from the previous optimization.

In particular, the optimized outer hexagon path is used as a mask for cropping the search image and further remove the background, improving the following GA search. Then, from the optimized inner hexagon, the diagonal on which the 8.8 logo lies is detected (i.e., "ax" in Figure 7b,c, found by summing the pixels intersecting the three diagonals and seeking the maximum), and the information is used as input for the last GA optimization.

It is important to point out that using a parametric template-mask defined on a continuous search space and implementing a GA optimization of the match between the corresponding discrete template image and the search image, it is possible to obtain a parametric estimation that goes beyond the pixel grid, leading to a super-resolution (i.e., similar to what obtained in [30]).

2.4.2. Objective Function

The change of paradigm from traditional TM to GA-adaptive TM is related to the use of GA for the estimation of the optimal parameters maximizing the match of the parametric template to the search image. In order to evaluate "how good" a reconstructed template is (N.B., reconstructed on the basis of the selected parameters), an objective function (commonly called utility function when referred to maximization problems or cost function when dealing with minimizations) is needed. In order to keep the link between TM and GA-adaptive TM, a possible utility function is the correlation function (i.e., *r* in Equation (11)). Nevertheless, in the literature, other commonly found functions are the Sum of Absolute Differences (i.e., SAD) or the Sum of Squared Differences (i.e., SSD), usually implemented as cost functions for a minimization problem.

In order to select the best objective function for this particular implementation, two considerations are fundamental. First, in this work the template is reconstructed to the same size ($J \times K = 191 \times 191$) of the search image so that all the objective functions can be easily implemented as:

$$r(param) = \sum_{j=1}^{J} \sum_{k=1}^{K} t[j,k|param]s[j,k] \tag{11}$$

$$SAD(param) = \sum_{j=1}^{J} \sum_{k=1}^{K} |t[j,k|param] - s[j,k]| \tag{12}$$

$$SSD(param) = \sum_{j=1}^{J} \sum_{k=1}^{K} \left(t[j,k|param] - s[j,k] \right)^2 \tag{13}$$

where $t[j,k|param]$ is the reconstructed template as a function of the corresponding parameters (i.e., *param*, see Figure 7) and $s[j,k]$ is the search image (e.g., a frame processed to obtain the result in Figure 4c).

Second, the t and s are binary, so that the possible results for single pixel information can be summarized as in Table 2:

Table 2. Correlation r, Sum of Absolute Differences (SAD) and Sum of Squared Differences (SSD) comparison for binary images.

t	s	$\rightarrow$	r	*SAD*	*SSD*
1	1		1	0	0
1	0		0	1	1
0	1		0	1	1
0	0		0	0	0

From Table 2, it is clear that SAD and SSD are equivalent in the case of binary images. Another relevant consideration regards the fact that correlation can be used for maximizing the match (i.e., the similarity), while SAD and SSD are suitable for minimization of the mismatch (i.e., the difference). Nevertheless, correlation rewards the similarity of white pixels (i.e., the 1) only, but neglects the black pixels (i.e., the 0). On the contrary, SAD and SSD penalize only the different pixels, or in other words, rewards both the white matching and the black matching pixels. As a result of this, the correlation was used as a utility function for the first GA (so that, thanks to the selected template shape Figure 7a, the 8.8 and JD logos are not accounted in the match), while the SAD was selected as a cost function for the second and third GA optimization.

2.4.3. Genetic Algorithm Optimization

Optimization is the selection of the best element from a set of available alternatives according to some criteria. In a more formal way, given an objective function $f : S \rightarrow R$ which links the search space of feasible solutions to the corresponding utility or cost, the optimization process seeks to find the element $x_0 \in S$ such that $f(x_0) \leq f(x) \; \forall \, x \in S$ (minimization) or such that $f(x_0) \geq f(x) \; \forall \, x \in S$ (maximization). Fixing a target for convenience, in the simplest case, an optimization problem corresponds to the minimization of a cost function over a search space obtained by constraining the overall Euclidean space. Or, $\underset{x \in S}{\operatorname{argmin}} f(x)$. From a mathematical point of view, the minimization of a function typically involves derivatives. Then, the more a function is complex (e.g., defined on a wide multidimensional support, non-continuous, or with non-continuous derivatives, featuring many local minima, etc.), the harder is the computation of such derivatives, so that the optimization may become very tricky in practical cases. Furthermore, the optimization is very likely to get stuck into local minima in the vicinity of an initial guess value for the optimum location (local optimization), with no guarantees (unless particular properties of the cost function i.e., convexity) that the result corresponds to the actual global minimum (global minimization).

In general, the assessment of the performance of an optimizer can be expressed in terms of:

- Exploration: the optimizer discovers a wide region of the search space,
- Exploitation: the optimizer "pounds the pavement" on a limited but promising region,
- Reliability: repeatability of the fund solution.

It is important to highlight that exploration and exploitation are competing properties. Local optimizers show very good exploitation at the expense of a very poor exploration. On the contrary, a good global optimizer should sacrifice exploitation to gain in exploration and speed. This is usually obtained taking advantage of heuristic or meta-heuristic techniques implementing some form of stochastic optimization.

An important category of global population-based metaheuristic optimization algorithms is the Evolutionary. An evolutionary algorithm (EA) uses mechanisms inspired by biological evolution, such as reproduction, mutation, recombination, and selection. Candidate solutions to the optimization problem play the role of individuals in a population, and the cost function determines the quality of a solution. A "direct search" is performed to find the best individuals within the population according to their quality. These best individuals are then selected to determine the offspring, namely the new trial solutions, which will substitute lower quality individuals.

The most famous EA is the Genetic Algorithm, developed by John Holland introduced genetic algorithms in 1960 based on the concept of Darwin's theory of evolution. The GA evolutionary cycle starts initiating a population randomly and evaluating the quality of each individual on the basis of his genotype. The best individuals are then selected to produce via modification the new offspring, while the worst are discarded. Modifications are stochastically triggered operators such as the crossover (the offspring is a random mix of the genotypes of their parents) or the mutation (the offspring features new genes which were not present in the parents). The first is important to ensure exploitation, while the second guarantees exploration of the search space of all possible genotypes. Finally, a new population is ready for starting again the cycle until some stopping criteria are met. The cycle is outlined in Figure 8.

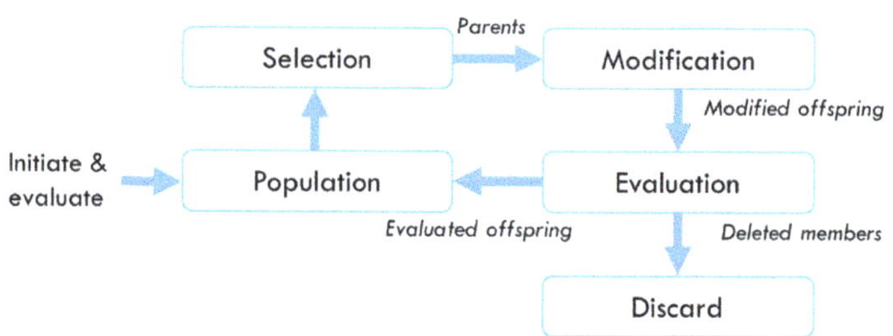

Figure 8. The GA evolutionary cycle.

The GA selected inputs in this work were:

- Population Size: $N_p = 100$.
- Elite Count: 5%. It defines the number of best individuals selected as a percentage of N_p.
- Crossover Fraction: 80%. It defines the offspring quantity at the next generation as a percentage of N_p. As the total N_p is fixed, the percentage of discarded individuals equals the crossover fraction.
- Default mutation: Shrinking Gaussian. Each newborn features a degree of random mutation which decreases in time according to the linear law: $\sigma_g = \sigma_{g-1}\left(1 - c\frac{g}{G}\right)$. Where $\sigma_0 = 1, c = 1$, and g is the generation index, increasing with time.
- Stopping criterion: maximum number of generations $G = 40$.

2.5. Overall Methodology

Finally, the overall methodology to be repeated for all the frames in the video is summarized in the following steps:

1. GA optimization of the outer hexagon template (Figure 7a) to match the search image (e.g., Figure 4c).

    ○   The outer hexagon path is used to make a mask isolating the foreground of interest and improving the next step.

2. GA optimization of the inner hexagon template (Figure 7b) to match the search image cropped using the outer hexagon path as mask.

    ○   The inner hexagon path is used to make a mask for isolating the foreground of interest and improving the next step.
    ○   The three inner hexagon diagonals are tested to find the diagonal around which the 8.8 logo is reported.

3. GA optimization of the 8.8 logo (Figure 7c) to match the search image cropped using the inner hexagon path as a mask.

Thanks to this procedure, all the 13 parameters of interest (Figure 7) can be estimated in all the frames of the video. The position of the 8.8 logo is used to identify a unique vertex of the inner hexagon, from which it is easy to derive the angular position of the screw (and then the angular position of the fan) over time.

IAS Estimation

Once the 13 parameters of interest are available over time, several different IAS estimations can be obtained. In this work, two methodologies are compared.

The first and simplest consists in differencing the angle over time and rescaling over the time interval determined by the sampling frequency (i.e., 30 $fps \rightarrow f_s = 30\ Hz$). Being $\alpha(n)$ is the angle of the vertex identified by the 8.8, and $t = n\,\Delta t = n/f_s$, it is easy to write:

$$\omega(t) = \frac{d\alpha}{dt} \cong \frac{\Delta\alpha}{\Delta t}\ [rad/s] \tag{14}$$

For this to be accurate, the angle signal should first be "unwrapped" (i.e., corrected by adding multiples of $\pm 2\pi$ when absolute jumps between consecutive elements are greater than π radians). Furthermore, a perfect recognition of all the three templates is required. Nevertheless, even if the match for two hexagons is good for all the frames, it is not the same for the 8.8 logo, which, in some cases (often at a determined angle because of illumination issues), is confounded with the manufacturer logo "JD" (e.g., see Figure 6). In this case, an error of about $\pm\pi$ radians should be compensated in a pre-processing stage.

The second method is based on phase demodulation via the Hilbert analytic signal of a fan speed-related harmonic [7,56]. In this particular case, it was noticed that the signal corresponding to the coordinates of the center of the outer hexagon (i.e., X_c, Y_c in Figure 7a) features a speed-related harmonic because of the not perfect alignment of the optical axis of the camera with the fan axis, leading to a circular movement of the tracked center point.

In a mathematical framework:

1. The analytic signal is computed via Hilbert transform

$$\alpha_{an}(t) = x(t) + iy(t) = A(n)e^{i\Phi(t)} \tag{15}$$

2. The instantaneous frequency is defined as

$$f(t) = \frac{1}{2\pi}\frac{d\Phi(t)}{dt} \quad [Hz] \tag{16}$$

3. From which, a more suitable discrete-time ($t = n\,\Delta t$) implementation can be derived [56]

$$f(n) = \frac{fs}{2\pi}\mathrm{atan}\left(\frac{x(n)y(n+1) - x(n+1)y(n)}{x(n)x(n+1) + y(n)y(n+1)}\right) \tag{17}$$

Despite the "unwrap" and the error compensation, the IAS estimated from the two methods could still be noisy, with large observable discontinuities which are unphysical and incompatible with the principle "Natura non facit saltus", so that a smoothing could be required.

Both the methods were tested on the data to validate the two IAS estimates.

3. Results

The application of the here described GA-adaptive TM led to the estimation of a set of 13 parameters fully specifying the selected template model over time (i.e., frame after frame).

As an example, the first two parameters obtained after the first GA step are the coordinates of the outer hexagon in the search space. These are reported in Figure 9. As can be seen from the picture, the X_c coordinate shows some jumps. These can be put in relation with the autofocus. In the same way, it is easy to notice that the coordinate Y_c is subject to a variation of the first-order statistics (i.e., the mean value) over time. This is highlighted by the estimation of the moving average (order: 50 samples) reported in yellow in the graph.

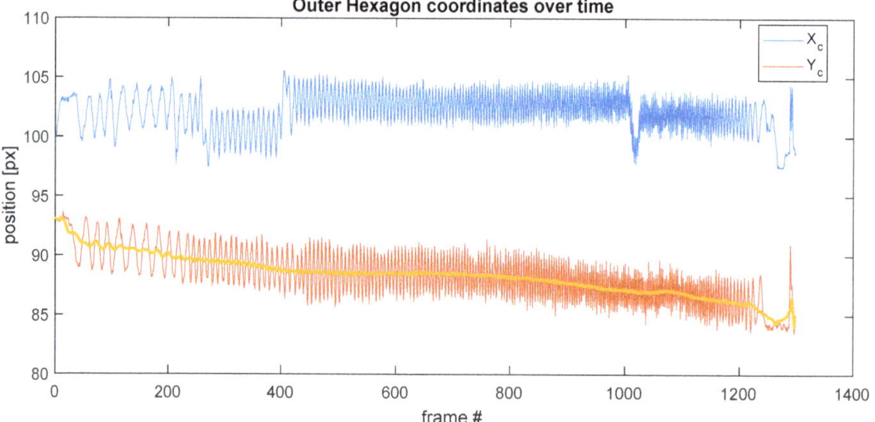

Figure 9. Outer hexagon coordinates over time (frame index). The moving average of the estimated parameter Y_c is added in yellow. N.B., The selected y coordinate ranges from the top to the bottom of the displayed image (e.g., in Figure 6), differently from the standard.

Similar graphs are available for all the 13 optimized parameters. The position of the inner hexagon center and the 8.8 logo vertex were then used to compute the angular position of the fan over time. As can be seen in Figure 10, the angle increments over time are quite uniform if the green, red and magenta points are not considered. In order to get a more consistent estimate, these points should be compensated. As introduced, the green points (360° error) can be avoided by performing and "unwrap" of the angle signal, while the red and magenta points correspond to a wrong localization of the 8.8 logo, which is confounded with the JD logo, so that the 180° error can be distinguished and compensated.

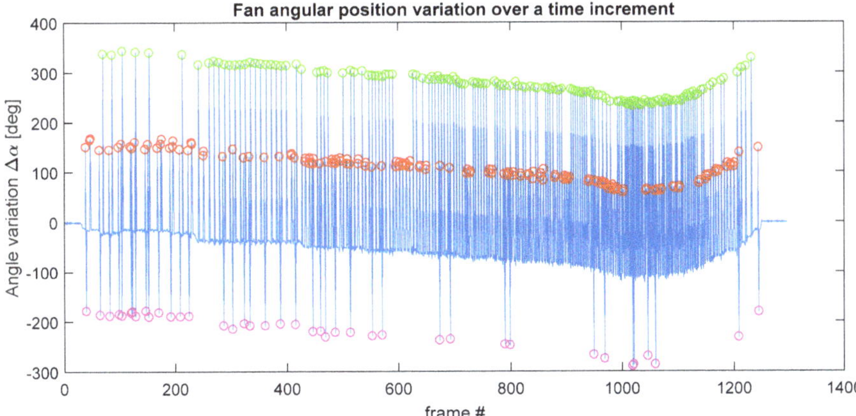

Figure 10. Fan angular position variation (angular increments $\Delta\alpha$) over time (frame index). In green are the points for the "unwrap", in red and magenta are the errors to be compensated (8.8 logo missed).

From the compensated angular increments over time, $\Delta\alpha$, it is easy to obtain the IAS estimate by normalizing this signal with respect to the constant $\Delta t = 1/f_s$, as indicated in Equation (14).

For the sake of comparison, Hilbert phase demodulation was performed on the running-mean-removed Y_c signal so as to produce a second IAS estimate (Equation (17)). The two estimates were then lowpass filtered to produce a more physically reasonable result. The IAS signals after a FIR1 lowpass filter (order: 50 samples, cutoff: $0,1\ f_s/2$) is reported in Figure 11.

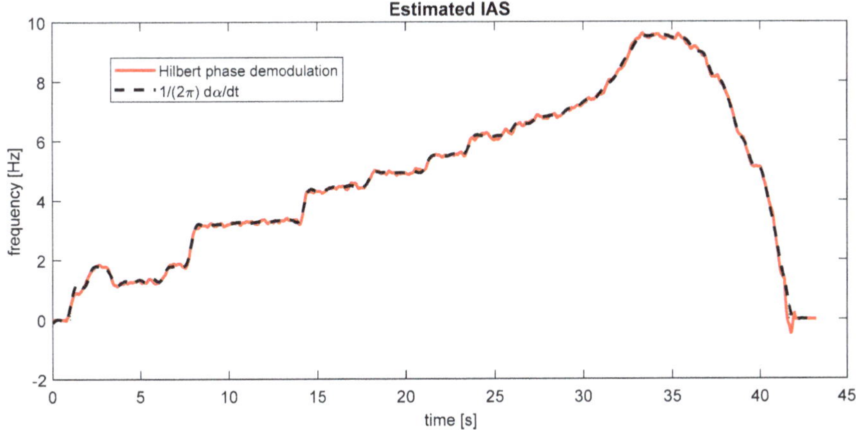

Figure 11. Final Instantaneous Angular Speed (IAS) estimates from the SURVISHNO video.

As can be noticed in Figure 11, the two methods lead to overlapping results. This, then, increases the confidence in the reliability and accuracy of the proposed video-tachometer procedure for the IAS estimation.

4. Discussion and Conclusions

The paper presented a novel method for implementing a cost-effective video-tachometer through a GA-adaptive Template Matching. The target was an offline implementation, as the proposed algorithm is not optimized enough and results slow if running on nowadays PCs. To give an idea, the software execution takes, per each frame, about 2s for the first GA, 2s for the second GA, and 10s for the third GA when using MATLAB R2018b on a machine with 8 GB of ram and an INTEL i7-7700 CPU at 3,60 GHz. Clearly, for obtaining just the Hilbert demodulation estimate of the IAS, only the first optimization step is needed, so that the computational burden can be strongly reduced, but it is not enough for real-time implementation.

In any case, the method proved to be effective in estimating the IAS of the fan despite the limits of the SURVISHNO 2019 video, acquired using a mobile phone. In fact, the wise selection of the search space effectively dealt with spatial aliasing (i.e., the rolling shutter effect), temporal aliasing (i.e., because of the 10 equal blades of the fan combined with a 30-fps sampling rate), and the additional autofocus distortions. Furthermore, the GA-adaptive implementation of TM reconstructing the binary edge template from a geometrical parametric model demonstrated its robustness to illumination changes and noise in general, as well as to rigid and non-rigid transformations. The issue of background changes and clutter was also tackled both in a pre-processing stage, by cropping and binarizing the search frame, and in the three-step GA optimization, exploiting the information from the previous stages for further focusing the TM on a smaller region of interest.

The here described GA-adaptive TM has the great advantage of allowing the localization of the template with a super-resolution that goes beyond the pixel-grid discretization. This allowed obtaining a robust and reliable estimate of the IAS, avoiding the need of expensive high-resolution encoders or tachometers, which are otherwise necessary for non-stationary machine diagnostics. When the speed is variable, in fact, the machine signature can only be highlighted by resampling the signal synchronously with the angular position of a reference shaft (i.e., performing the so-called Computed Order Tracking to get to the order domain). This way, the events which are phase-locked to the reference shaft (e.g., the intake of a 4 strokes diesel engine to the crankshaft, or the meshing of a broken tooth of a gearwheel to the supporting shaft, etc.) are put in evidence. Furthermore, the IAS is considered a precious diagnostic information per se, so that the analysis of IAS anomalies is spreading in the field of condition monitoring.

To conclude, given the here underlined strengths, the proposed signal processing gives an effective and reliable tool able to foster the IAS-based condition monitoring, setting the state of the art for video-tachometric acquisitions.

Supplementary Materials: The video was downloaded online during the international conference SURVISHNO 2019 at https://survishno.sciencesconf.org/resource/page/id/18.

Author Contributions: Conceptualization, A.P.D.; data curation, A.P.D.; formal analysis, A.P.D.; funding acquisition, L.G.; investigation, A.P.D.; methodology, A.P.D.; project administration, L.G.; software, A.P.D.; supervision, L.G.; validation, L.G.; visualization, A.P.D.; writing—original draft, A.P.D.; writing—review and editing, A.P.D. All authors have read and agreed to the published version of the manuscript.

Funding: This research received no external funding.

Acknowledgments: We thank the SURVISNO 2019 organization committee and in particular Quentin LECLERE, Hugo ANDRE and Emmanuel FOLTETE for acquiring the video and organizing the challenge.

Conflicts of Interest: The authors declare no conflict of interest.

References

1. Lin, J.; Zhao, M. A review and strategy for the diagnosis of speed-varying machinery. In Proceedings of the 2014 International Conference on Prognostics and Health Management, Cheney, WA, USA, 22–25 June 2014.
2. Fyfe, K.R.; Munck, E.D.S. Analysis of computed order tracking. *MSSP* **1997**. [CrossRef]

3. Brandt, A.; Lago, T.; Ahlin, K.; Tuma, J. Main principles and limitations of current order tracking methods. *Sound Vib.* **2005**, *39*, 19–22.

4. Bonnardot, F.; El Badaoui, M.; Randall, R.B.; Daniere, J.; Guillet, F. Use of the acceleration signal of a gearbox in order to perform angular resampling (with limited speed fluctuation). *MSSP* **2005**. [CrossRef]

5. Coats, M.; Sawalhi, N.; Randall, R.B. Extraction of tacho information from a vibration signal for improved synchronous averaging. *Acoustics* **2009**.

6. Remond, D.; Antoni, J.; Randall, R.B. Instantaneous Angular Speed (IAS) processing and related angular applications. *MSSP* **2014**, *45*, 24–27. [CrossRef]

7. Peeters, C.; Leclère, Q.; Antoni, J.; Lindahl, P.; Donnal, J.; Leeb, S.; Helsen, J. Review and comparison of tacholess instantaneous speed estimation methods on experimental vibration data. *MSSP* **2019**. [CrossRef]

8. Antoni, J.; Griffaton, J.; André, H.; Avendaño-Valencia, L.D.; Bonnardot, F.; Cardona-Morales, O.; Castellanos-Dominguez, G.; Daga, A.P.; Leclère, Q.; Vicuña, C.M.; et al. Feedback on the surveillance 8 challenge: vibration-based diagnosis of a safran aircraft engine. *MSSP* **2017**. [CrossRef]

9. Ben Sasi, A.Y.; Gu, F.; Payne, B.; Ball, A. Instantaneous angular speed monitoring of electric motors. *J. Qual. Maint. Eng.* **2004**. [CrossRef]

10. Lebaroud, A.; Clerc, G. Diagnosis of Induction Motor Faults Using Instantaneous Frequency Signature Analysis. *ICEM* **2008**. [CrossRef]

11. Spagnol, M.; Bregant, L.; Boscarol, A. Electrical Induction Motor Higher Harmonics Analysis Based on Instantaneous Angular Speed Measurement. In Proceedings of the Fourth International Conference on Condition Monitoring of Machinery in Non-Stationary Operations, CMMNO'2014, Lyon, France, 15–17 December 2014. [CrossRef]

12. Yang, J.; Pu, L.; Wang, Z.; Zhou, Y.; Yan, X. Fault detection in a diesel engine by analysing the instantaneous angular speed. *MSSP* **2001**. [CrossRef]

13. Desbazeille, M.; Randall, R.B.; Guillet, F.; El Badaoui, M.; Hoisnard, C. Model-based diagnosis of large diesel engines based on angular speed variations of the crankshaft. *MSSP* **2010**. [CrossRef]

14. Chang, Y.; Hu, Y. Monitoring and fault diagnosis system for the diesel engine based on instantaneous speed. *ICCAE Conf.* **2010**. [CrossRef]

15. Antonopoulos, A.K.; Hountalas, D.T. Effect of instantaneous rotational speed on the analysis of measured diesel engine cylinder pressure data. *Energy Convers Manag.* **2012**. [CrossRef]

16. Lin, T.; Tan, A.; Ma, L.; Mathew, J. Condition monitoring and fault diagnosis of diesel engines using instantaneous angular speed analysis. *Proc. Inst. Mech. Eng. Part C: J. Mech. Eng. Sci.* **2014**. [CrossRef]

17. Liu, Y.; Gu, L.; Yang, B.; Wang, S.; Yuan, H. A new evaluation method on hydraulic system using the instantaneous speed fluctuation of hydraulic motor. *Proc. Inst. Mech. Eng. Part C: J. Mech. Eng. Sci.* **2018**. [CrossRef]

18. Roy, S.K.; Mohanty, A.R.; Kumar, C.S. Fault detection in a multistage gearbox by time synchronous averaging of the instantaneous angular speed. *J. Vib. Control* **2016**. [CrossRef]

19. Li, B.; Zhang, X. A new strategy of instantaneous angular speed extraction and its application to multistage gearbox fault diagnosis. *J. Sound Vib.* **2017**. [CrossRef]

20. Zhao, M.; Jia, X.; Lin, J.; Lei, Y.; Lee, J. Instantaneous speed jitter detection via encoder signal and its application for the diagnosis of planetary gearbox. *MSSP* **2018**. [CrossRef]

21. Liang, L.; Liu, F.; Kong, X.; Li, M.; Xu, G. Application of Instantaneous Rotational Speed to Detect Gearbox Faults Based on Double Encoders. *Chin. J. Mech. Eng.* **2019**. [CrossRef]

22. Zimroz, R.; Urbanek, J.; Barszcz, T.; Bartelmus, W.; Millioz, F.; Martin, N. Measurement of Instantaneous Shaft Speed by Advanced Vibration Signal Processing—Application to Wind Turbine Gearbox. *Metrol. Meas. Syst.* **2011**. [CrossRef]

23. Renaudin, L.; Bonnardot, F.; Musy, O.; Doray, J.B.; Rémond, D. Natural roller bearing fault detection by angular measurement of true instantaneous angular speed. *MSSP* **2010**. [CrossRef]

24. Bourdon, A.; André, H.; Rémond, D. Introducing angularly periodic disturbances in dynamic models of rotating systems under non-stationary conditions. *MSSP* **2014**. [CrossRef]

25. Lamraoui, M.; Thomas, M.; El Badaoui, M.; Girardin, F. Indicators for monitoring chatter in milling based on instantaneous angular speeds. *MSSP* **2014**. [CrossRef]

26. Leclere, Q.; André, H.; Antoni, J. A multi-order probabilistic approach for Instantaneous Angular Speed tracking. debriefing of the CMMNO'14 diagnosis contest. *MSSP* **2016**. [CrossRef]

27. Li, Y.; Gu, F.; Harris, G.; Ball, A.; Bennett, N.; Travis, K. The measurement of instantaneous angular speed. *MSSP* **2005**. [CrossRef]

28. Sushma Sri, M. A study on motion estimation. *Int. J. Mod. Trends Eng. Res. (IJMTER)* **2017**. [CrossRef]

29. Torr, P.H.S.; Zisserman, A. Feature Based Methods for Structure and Motion Estimation. *ICCV Workshop Vis. Algorithms* **1999**. [CrossRef]

30. Irani, M.; Anandan, P. About Direct Methods. *ICCV Workshop Vis. Algorithms* **1999**. [CrossRef]

31. Egmont-Petersen, M.; de Ridder, D.; Handels, H. Image processing with neural networks—A review. *Pattern Recognit.* **2002**, *35*, 2279–2301. [CrossRef]

32. Tolba, A.; El-Baz, A.; El-Harby, A. Face Recognition: A Literature Review. *Int. J. Signal Process.* **2005**, *2*, 88–103.

33. Solanki, K.; Pittalia, P. Review of Face Recognition Techniques. *Int. J. Comput. Appl.* **2016**, *133*, 20–24. [CrossRef]

34. Brunelli, R. *Template Matching Techniques in Computer Vision: Theory and Practice*; Wiley: Hoboken, NJ, USA, 2009.

35. Leavers, V.F. Computer Vision: Shape Detection. In *Shape Detection in Computer Vision Using the Hough Transform*; Springer: London, UK, 1992. [CrossRef]

36. Talmi, I.; Mechrez, R.; Zelnik-Manor, L. Template Matching with Deformable Diversity Similarity. Available online: http://openaccess.thecvf.com/content_cvpr_2017/html/Talmi_Template_Matching_With_CVPR_2017_paper.html (accessed on 20 December 2019).

37. Lee, H.; Kwon, H.; Robinson, R.M.; Nothwang, W.D. DTM: Deformable Template Matching. In Proceedings of the 2016 IEEE International Conference on Acoustics, Speech and Signal Processing (ICASSP), Shanghai, China, 20–25 March 2016.

38. Muramatsu, S. Strategy of high-speed template matching and its optimization by using GA. *Syst. Comput. Jpn.* **2003**. [CrossRef]

39. Karungaru, S.; Fukumi, M.; Akamatsu, N. Face recognition using genetic algorithm based template matching. *IEEE Int. Symp. Commun. Inf. Technol.* **2004**. [CrossRef]

40. Velasco, H.M.G.; Orellana, C.J.G.; Macías, M.M.; Caballero, R.G.; Sotoca, M.I.A. Application of Repeated GA to Deformable Template Matching in Cattle Images. In Proceedings of the Third International Workshop, AMDO 2004, Palma de Mallorca, Spain, 22–24 September 2004. [CrossRef]

41. Zhang, X.; Liu, Q.; Li, M. A Template Matching Method Based on Genetic Algorithm Optimization. *Appl. Mech. Mater.* **2012**. [CrossRef]

42. El-Baz, A.; Elnakib, A.; Abou-El-Ghar, M.; Gimel'farb, G.; Falk, R.; Farag, A. Automatic Detection of 2D and 3D Lung Nodules in Chest Spiral CT Scans. *Int. J. Biomed. Imaging* **2013**. [CrossRef]

43. Kay, S.; Rangaswamy, M. The Ubiquitous Matched Filter: A Tutorial and Application to Radar Detection. In *Classical, Semi-classical and Quantum Noise*; Cohen, L., Poor, H., Scully, M., Eds.; Springer: New York, NY, USA, 2012. [CrossRef]

44. Turin G., L. An introduction to matched filters. *IRE Trans. Inf. Theory* **1960**, *6*, 311–329. [CrossRef]

45. Vleck, J.V.; Middleton, D. A theoretical comparison of the visual, aural, and meter reception of pulsed signals in the presence of noise. *J. Appl. Phys.* **1946**, *17*, 940–971. [CrossRef]

46. North, D.O. An analysis of the factors which determine signal/noise discrimination in pulsed-carrier systems. *Proc. IEEE* **1943**, *51*, 1016–1027. [CrossRef]

47. Mohd, F.Z.; Hoo, S.C.; Shahrel, A.S. Object Shape Recognition in Image for Machine Vision Application. *Int. J. Comput. Theory Eng.* **2012**, 76–80. [CrossRef]

48. Kumar, V.; Pandey, S.; Pal, A.; Sharma, S. Edge Detection Based Shape Identification. Available online: https://arxiv.org/ftp/arxiv/papers/1604/1604.02030.pdf (accessed on 20 December 2019).

49. Chai, T.Y.; Rizon, M.; San, W.; Seong, T. Facial Features for Template Matching Based Face Recognition. *Am. J. Appl. Sci.* **2009**. [CrossRef]

50. ITU-R recommendation BT.601 standard, revision 7. 2011. Available online: https://www.itu.int/rec/R-REC-BT.601/en (accessed on 20 December 2019).

51. Otsu, N. A Threshold Selection Method from Gray-Level Histograms. *IEEE Trans. Syst. Man Cybern.* **1979**, *9*, 62–66. [CrossRef]

52. Bradley, D.; Roth, G. Adaptive Thresholding using the Integral Image. *J. Graph. Tools* **2007**. [CrossRef]

53. MATLAB Matworks "imbinarize" function. Available online: https://www.mathworks.com/help/images/ref/imbinarize.html (accessed on 20 December 2019).
54. Sobel, I.; Feldman, G. An Isotropic 3x3 Image Gradient Operator. Available online: https://www.researchgate.net/publication/239398674_An_Isotropic_3x3_Image_Gradient_Operator (accessed on 20 December 2019).
55. MATLAB Matworks "edge" function. Available online: https://it.mathworks.com/help/images/ref/edge.html (accessed on 20 December 2019).
56. Barnes, A.E. The calculation of instantaneous frequency and instantaneous bandwidth. *Geophysics* **1992**, *57*, 1396–1524. [CrossRef]

Article

Blended Filter-Based Detection for Thruster Valve Failure and Control Recovery Evaluation for RLV

Hongqiang Sun and Shuguang Zhang *

School of Transportation Science and Engineering, Beihang University, Beijing 100191, China;
hqsun@buaa.edu.cn
* Correspondence: gnahz@buaa.edu.cn

Received: 23 August 2019; Accepted: 29 October 2019; Published: 1 November 2019

Abstract: Security enhancement and cost reduction have become crucial goals for second-generation reusable launch vehicles (RLV). The thruster is an important actuator for an RLV, and its control normally requires a valve capable of high-frequency operation, which may lead to excessive wear or failure of the thruster valve. This paper aims at developing a thruster fault detection method that can deal with the thruster fault caused by the failure of the thruster valve and play an emergency role in the cases of hardware sensor failure. Firstly, the failure mechanism of the thruster was analyzed and modeled. Then, thruster fault detection was employed by introducing an angular velocity signal, using a blended filter, and determining an isolation threshold. In addition, to support the redundancy management of the thruster, an evaluation method of the nonlinear model-based numerical control prediction was proposed to evaluate whether the remaining fault-free thruster can track the attitude control response performance under the failure of the thruster valve. The simulation results showed that the method is stable and allowed for the effective detection of thruster faults and timely evaluation of recovery performance.

Keywords: reusable launch vehicle; thruster valve failure; thruster fault detection; Kalman filter

---

1. Introduction

The successful flight of Colombia in 1981, the first space shuttle of the United States, pioneered the partial reuse of a space launch vehicle [1]. However, it had high launch costs and a long preparation period. Applying the learning experience from this shuttle mission [2], NASA implemented the second-generation reusable launch vehicle (RLV) research program. At present, X-37B has been successfully launched several times, with the longest orbiting flight time of 718 days [3], demonstrating a breakthrough in RLV technology.

As a new type of aerospace flight vehicle, the effects of system failure caused by various actuators, sensors, or system components are of more concern to researchers. Statistics from the Federal Aviation Authority (FAA) and the National Transport Safety Board (NTSB) show that among the flight accidents over the past decade, hardware and system failure were the main cause of loss-of-control during flight [4]. Considering the harsh aerospace environment and the complexity of aerospace tasks, there is a call for intelligent fault tolerance control that can handle component faults and does not increase hardware complexity, cost, and mass [5].

RLV reentry usually relies on the thruster to supplement the aerodynamic actuator for attitude control. Specifically, during orbit, deorbit, and initial reentry phase, the thruster is the main actuator for the attitude control of the RLV.

Currently, the pulse width modulation (PWM) method [6,7] is mainly used for thruster control in spacecraft. When a duty cycle of a thruster is modulated using this method, the thruster can be considered as a continuous actuator. Its advantage lies in the application of the continuous multivariable control method to the RLV's attitude control. However, the PWM method requires a thruster that is capable of high-frequency operation, which may lead to excessive wear of the thruster valve. In addition, the PWM duty cycle of the thruster system is subject to a certain practical limit [8].

One common approach to detect and isolate spacecraft thruster faults is to install pressure, temperature, and electrical sensors on the thrusters. The use of these additional sensors makes the fault detection and isolation quite simple and robust, as these sensors can assist in the direct detection of thrust. However, the need for additional sensors adds to the complexity, mass, and volume requirements of the spacecraft. This type of system is used, for example, on NASA's space shuttle orbiter [9]. When such extensive sensing is unavailable, an online fault diagnosis approach for thrusters can play an emergency role in the cases of hardware sensor failure or work as a supplementary reference for online monitoring [10].

A maximum-likelihood-based approach was proposed in the literature [11,12] to detect leaking thrusters for the space shuttle orbiter's reactive control system jets; the soft failure was used to designate a fuel or oxidizer leak in a vehicle reactive control system jet. For the fuel leak, it only described the state in which one valve was open and the jet did not fire, but it did not consider the fuel leak after valve wear and other valve failure cases. In the literature [13,14], fault detection and isolation methods were developed based on the exponentially weighted recursive least squares estimation using an accelerometer and angular rate sensors. A neural network then provided adaptive control reconfiguration to multiple destabilizing hard and soft thruster faults. However they did not provide an analysis of the failure caused by thruster valves, or, more particularly, the impact of thruster failure caused by valve performance degradation on the RLV or spacecraft. A nonlinear fault detection observer was designed in literature [15] to realize active fault tolerance tracking and RLV attitude control, combined with adaptive sliding mode technology, by the actuator redundancy. However, the fault detection observer was assumed to be idealized and less affected by the sensor noise and disturbances. In the literature [16], a velocity-free nonlinear proportional-integral control allocation scheme was provided for the fault-tolerant attitude control of flexible spacecraft under thruster redundancy. A uniform actuator fault model was established, but the system states before and after the fault were not considered; moreover, the noise signals that approximate the control input were also not regarded, leading to an output error from the model. The literature [17] proposed a sliding mode observer with an equivalent output injection to reconstruct external disturbances and actuator faults and developed a sliding mode-based attitude controller using an exponential reaching law. The exponential reaching law relies on the state of the system and the corresponding formation of the estimated parameters, although in the theoretical design process, the authors described the partial failure of the thruster using the thruster fault modeling. However, it could not detect faults under low thrust conditions due to the thruster leakage in the simulation verification process. The simulation results also showed that the fault detection time frame of the designed sliding observer was more than 6 s. In literature [18], an observer-based fault-tolerant controller was constructed such that the states of the resulting closed-loop systems were uniformly and ultimately bounded in the presence of model parameter uncertainty, external disturbances, time-varying input delay, actuator faults, and actuator saturation. However, the controller gain matrix and observer gain matrix were required to be designed for satisfying the results of the uniformly and ultimately bounded system. This method helped in dealing with several problems at the expense of optimal control performance in different situations, and the fixed gain of the observer did not consider the requirements of specific fault detection such as false alarm rates, fault detection rates, and fault detection time.

Literature [19] showed that any motion-based fault detection and isolation is reliant on the accuracy of the measurement data used, thus, great care has been taken here to obtain highly accurate angular and translational acceleration estimates from the gyros and accelerometers. The motion-based method presented in this work was used and extended in this research.

Unlike previous studies, in this study, the failure of thruster valves, including valve wear and leakage failure in addition to valve jams and difficulty in closure, was investigated. We proposed a detection solution for the failures based on motion state estimation with particular consideration of the noise that affects the RLV operation in initial reentry, on-orbit, and de-orbit phase, which is caused by electromagnetic radiation in near-Earth space [20]. In this paper, a physical filter was used in combination with a Kalman filter to suppress the input interference of the approximate control of the thruster valve, enhance the consistency of physical input and algorithm model input, improve the accuracy of motion state estimation, and thereby improve the fault detection rate.

The paper aims at developing an online thruster fault detection method that can handle thruster valve failure and does not increase more hardware complexity or cost.

Firstly, the failure of thruster valves under different working conditions was analyzed, and the thruster model caused by valve failure was established.

Then, on the basis of the thruster model, a motion-based blending filter fault detection method is proposed. The blending filter was used to estimate the thruster state and suppress the noise signals that approximate the control signal. The fault detection was conducted by comparing the variance-based function of the observed output in a certain time window with the threshold.

Finally, based on the thruster redundancy, a nonlinear model-based numerical control prediction method is proposed to evaluate whether the remaining fault-free thruster can track the attitude control response performance under thruster valve failure, and conducted for longitudinal attitude control as an example.

2. Blending Filter-Based Thruster Fault Detection

2.1. Equations and Model

Ignoring the installation gap error of the thruster, the following can be assumed:

- The RLV can be treated as an ideal rigid body and any elastic vibration can be ignored;
- The dynamic characteristics of the navigation system can be ignored, and the feedback state of the vehicle can be regarded as the ideal state.

For longitudinal attitude control, it can be assumed that $\beta \approx 0$, $\varnothing \approx 0$, and, ignoring the earth's rotation, the flight dynamics equations [21] can be written as follows:

$$
\begin{cases}
\dot{h} = V \sin(\theta - \alpha) \\
\dot{V} = -\dfrac{D}{m_b} - \dfrac{\mu \sin(\theta - \alpha)}{r^2} \\
\dot{\alpha} = q - \dfrac{L}{m_b V} + \dfrac{(\mu - V^2 r)\cos(\theta - \alpha)}{V^2 r} \\
\dot{\theta} = q \\
\dot{q} = \dfrac{M}{I_y}
\end{cases}
\tag{1}
$$

where h is the altitude of the flight vehicle, V is the velocity, α is the angle of attack (AoA), θ is the pitch angle, q is the pitch rate, m_b is the mass of the flight vehicle, L and D are the lift force and drag force, respectively, μ is the product of the gravitational constant and the mass of the Earth, r is the distance from the flight vehicle to the center of gravity, I_y is the moment of inertia, and M is the pitch moment that is the input and is induced by the thruster.

The thruster valve is normally used to control the thrust from the thruster. When the thruster valve is open, the thrust produced by the thruster is not adjustable, so according to the thruster valve switching state, the fault-free thruster has only two states: normal-rated thrust and no thrust. The mathematical description of this can be written as follows:

$$
F_i =
\begin{cases}
0, U = 0 \\
F, U = 1
\end{cases}
\tag{2}
$$

where F is the normal rated thrust of the thruster, U is the switch command of the thruster valve that represents the control input, 0 is the command to turn off the thruster valve, and 1 is the command to turn on the thruster valve. In this paper, the control signal refers to the physical signal corresponding to the switch command of the thruster valve (control input).

2.2. Thruster Model Based on Valve Failure

Figure 1 shows the basic principle of the thruster. By supplying power to the coil, the electromagnet generates a magnetic force that drives the thruster valve to either switch on or off, so the fuel combines with the catalyst and, after heating, generates injection gas to provide reaction thrust to the RLV.

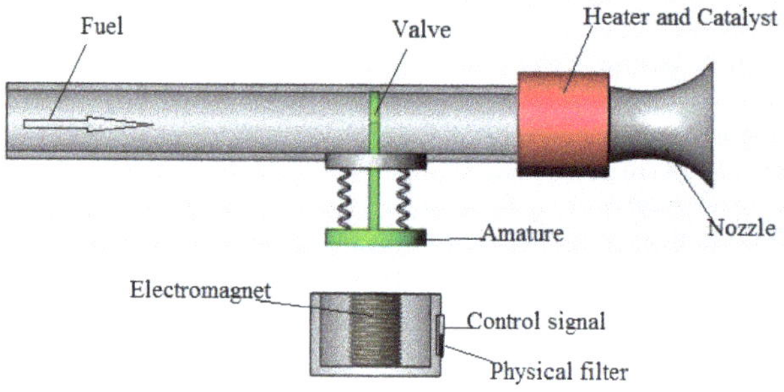

Figure 1. Structural diagram of the thruster.

For this study, the possible thruster faults caused by the failure conditions of the thruster valve are:

Fault A: The thruster valve is closed; a control signal is activated to turn on the valve, but the thruster valve does not open and a pipe jam fault occurs, that is:

$$F_i \equiv 0, when \begin{cases} U(k-1) = 0 \\ U(k) = 1 \end{cases}. \tag{3}$$

Fault B: The thruster valve is open; a control signal is applied to turn off the valve, but the thruster valve does not close and the pipeline cannot be closed, that is:

$$F_i \equiv F, when \begin{cases} U(k-1) = 1 \\ U(k) = 0 \end{cases}. \tag{4}$$

Fault C: The thruster valve is open; after a control signal is given, the valve-closing performance of the fuel pipe attenuates so that it is not completely executed in accordance with the control signal, and the valve only partially closes, that is:

$$\begin{cases} F_i(k-1) = F, \; U(k-1) = 1 \\ F_i(k) = \tau F, U(k) = 0 \end{cases}, \tag{5}$$

where τ refers to the ratio of the incomplete closing distance to the width of the nozzle after performance degradation.

2.3. Blending the Physical Filter and Kalman Filter for Fault Detection

The Kalman filter conducts timely calculations, has a small cache space, and estimates based on sensor feedback data. In this study, we used the aforementioned filter to estimate the attitude angular rate of the RLV.

The state variables $x \in \mathbb{R}^n$ of the discrete time process can be estimated by the Kalman filter and can be described by the following discrete random difference equations:

$$\begin{aligned}
\mathbf{x}_k &= \mathbf{A}\mathbf{x}_{k-1} + \mathbf{B}\mathbf{u}_{k-1} + \mathbf{w}_{k-1} \\
\mathbf{z}_k &= \mathbf{H}\mathbf{x}_k + \mathbf{v}_k
\end{aligned} \tag{6}$$

where the random signals $\mathbf{w}_{k-1}$ and $\mathbf{v}_k$ represent the process noise and observation noise, respectively. They can be assumed to be white noises and are independent of each other. Their probability density obeys the normal distribution of expectation 0, variance Q and expectation 0, variance R, respectively.

The Kalman filter comprises the prediction part based on the time series, and the correction and updating part based on the measurement feedback, wherein the prediction part is given by the following formulae:

$$\begin{aligned}
\hat{\mathbf{x}}_k^- &= \mathbf{A}\hat{\mathbf{x}}_{k-1} + \mathbf{B}\mathbf{u}_{k-1} \\
\mathbf{P}_k^- &= \mathbf{A}\mathbf{P}_{k-1}\mathbf{A}^T + \mathbf{Q}
\end{aligned} \tag{7}$$

where $\hat{\mathbf{x}}_k^-$ represents the forward-calculation state variables, $\mathbf{P}_k^-$ is the forward-calculation error covariance, and $\hat{\mathbf{x}}_{k-1}$ and $\mathbf{P}_{k-1}$ are the initial estimations.

The measurement feedback correction formulae are:

$$\begin{aligned}
\mathbf{K}_k &= \mathbf{P}_k^- \mathbf{H}^T (\mathbf{H}\mathbf{P}_k^- \mathbf{H}^T + \mathbf{R})^{-1} \\
\hat{\mathbf{x}}_k &= \hat{\mathbf{x}}_k^- + \mathbf{K}_k (\mathbf{z}_k - \mathbf{H}\hat{\mathbf{x}}_k^-) \\
\mathbf{P}_k &= (1 - \mathbf{K}_k \mathbf{H})\mathbf{P}_k^-
\end{aligned} \tag{8}$$

where $\mathbf{K}_k$ is the Kalman filter's gain, $\mathbf{z}_k$ represents the observation variables, $\hat{\mathbf{x}}_k$ is the updated estimate, and $\mathbf{P}_k$ is the updated error covariance.

In the longitudinal attitude control of a thruster, the short-cycle mode manipulation moment has a greater influence on the pitch rate. In this study, the pitch rate was used as a variable state of the fault detection and the Kalman filter was designed for the state of estimation, and combined with Equation (1). The equation of the state of the thruster longitudinal attitude fault detection can be expressed as follows:

$$q_t = q_{t-1} + \frac{M\Delta t}{I_y}, \tag{9}$$

where q_t is the pitch rate at the next moment, q_{t-1} is the pitch rate at the current time. The approximate interval time of the thruster control command is Δt. Equation (9) is obtained using the forward Euler method. The filter can be used for the estimation of one dimensional or multidimensional random process. In Equation (6) the pitch rate is the state variables, $\mathbf{A} = 1$, $\mathbf{B} = 1/I_y$, $\mathbf{H}$ is considered an ideal sensor gain of 1, and $\mathbf{u}_{k-1}$ refers to the pitch moment generated during the interval time of thruster control command. In the system design, the related allowable deviation range is given for the state variable and the observation owing to the uncertainty caused by noise. Therefore, the white noise variances Q and R can be set according to the deviation range of the state variable and the observation variable.

In this study, the state estimation of the Kalman filter is combined with the physical feedback and the prediction calculation. It is used as a software algorithm that predicts a follow-up state based on the model of a current state, and it is necessary to consider the consistency between the model input and the input of the physical system. If there is an approximate control signal noise in the input of the physical system, it may lead to the following: one is that the input of the physical system will deviate from the input of the model; the other is that the input of physical system has a great influence on the sensor data generated by the actual physical system, which is then introduced into the Kalman filter for state estimation. These challenges may result in the failure of the Kalman filter to estimate the state, as shown in Figure 2.

In the RLV flight, several thruster missions, including on-orbit, de-orbit, and initial reentry, are performed in the near-earth space environment. However, electromagnetic radiation in the near-earth space environment is very complex [22]. Furthermore, the noise caused by electromagnetic induction or charging effect of the vehicle can easily interfere with the control signal [23]. However, the vehicle control input usually works within a certain frequency band. For example, the flight attitude control command period is 20 ms, the guidance command period is 100 ms, and the no-command state still sends the reference pulse based on the command period. Therefore, in this study, a bandpass physical filter was introduced and applied before the control signal input in order to suppress the approximate control signal noise. The input position of the physical filter is shown in Figure 1. The approximate control signal noise outside the frequency band, which is caused by electromagnetic radiation in space, will be suppressed after passing through the physical filter. This ensures that the state estimation performed using the Kalman filter is more accurate. A sample circuit diagram of the physical filter is shown below in Figure 3.

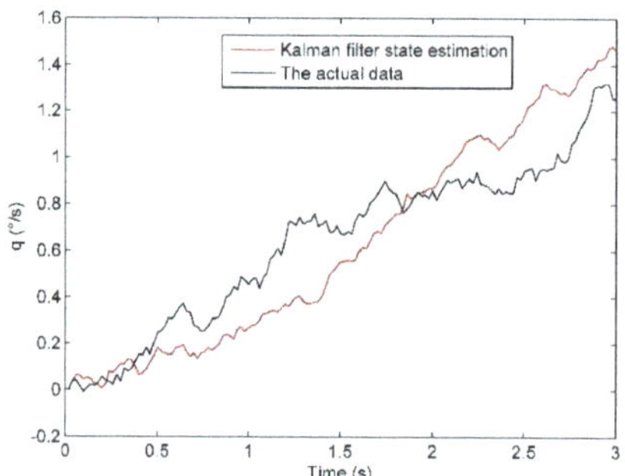

Figure 2. Kalman filter state estimation under the interference of approximate control signal noise.

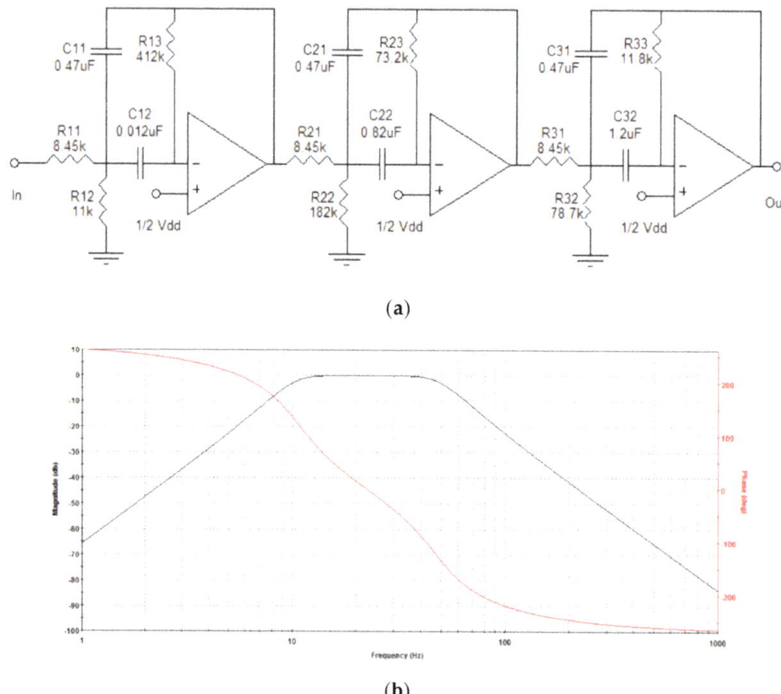

(a)

(b)

Figure 3. Example of a bandpass filter: (**a**) bandpass filter circuit diagram; (**b**) frequency response diagram.

The filter is designed using the FilterLab 2.0 software; the passband frequency is from 10 Hz to 50 Hz, which meets the control input period.

The introduction of the physical filter improved the input consistency between the model and the physical system as shown in the simulation results given in Figure 4. Blending the physical filter and Kalman filter results in a smaller state estimation bias in comparison to the Kalman filter.

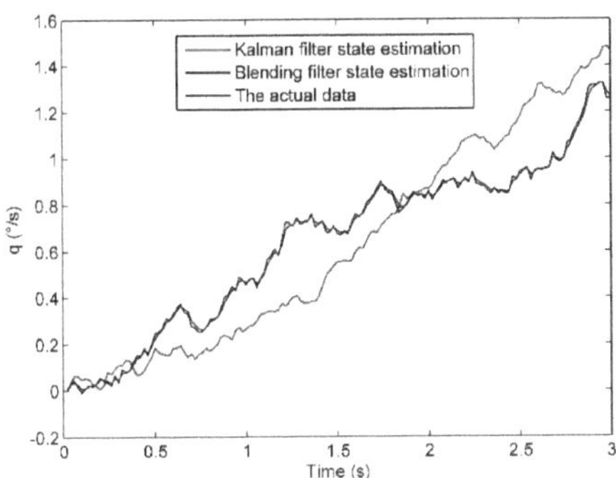

Figure 4. Blending of the physical filter and Kalman filter state estimation under the interference of approximate control signal noise.

In the attitude loop, the onboard calculation period is typically in the order of several tens of milliseconds. In this paper, the residual deviation between the predicted output of the blending filter ($\hat{\mathbf{z}}_k$) and the sensor measurement feedback ($\mathbf{z}_k$) was used as the observation output in each period for three types of faults, that is:

$$\begin{aligned} \mathbf{r}_k &= \mathbf{z}_k - \hat{\mathbf{z}}_k \\ \mathbf{r}_k &= \mathbf{H}\mathbf{x}_k + \mathbf{v}_k - \mathbf{H}\hat{\mathbf{x}}_k \end{aligned} \tag{10}$$

Fault-free condition: $E[\mathbf{r}_k] = 0$, their probability density obeys the normal distribution of expectation 0, variance R.

Fault condition: $E[\mathbf{r}_k] \neq 0$

A variance test is used for the residual evaluation for a selected time frame, and analysis; see Section 4.1.

Establishing the binary hypothesis:

H0: fault-free

H1: fault

The designed fault detection function (weighted sum of squared residuals) was:

$$\lambda_k = \mathbf{r}_k^T \mathbf{A}_k^{-1} \mathbf{r}_k, \tag{11}$$

where $\mathbf{A}_k = diag(\sigma_k^2)$, σ_k are the $\mathbf{r}_k$ group standard deviation, then $\lambda_k \sim \chi^2(m)$, m equals the dimension of $\mathbf{z}_k$.

The fault-free was determined by:

$$\lambda_k \leq T_D. \tag{12}$$

The fault was detected by:

$$\lambda_k > T_D. \tag{13}$$

To determine the T_D according to the requirement of the system false alarm rate P_F:

$$P_F = \alpha_T, \tag{14}$$

then,

$$P_F = \int_{T_D}^{\infty} \chi^2(m)d\lambda = \alpha_T \rightarrow T_D, \tag{15}$$

where α_T is the expected value.

3. Thruster Control Recovery Evaluation Based on Proposed Fault Detection

An RLV is a highly maneuverable vehicle that can implement missions of on-orbit, de-orbit, and reentry in the atmosphere. For different missions, the attitude control response needs to achieve different performances according to guidance commands, but when the working thrusters fail, the attitude control response performance will be affected. As an important actuator for an RLV, the thruster is usually designed with redundancy to ensure a safe flight. Based on the redundancy of the thruster, stopping the thruster failure and quickly selecting the fault-free thruster to recover the attitude control response performance are very helpful to improve the fault-tolerant ability of the RLV.

This section describes the nonlinear model-based numerical prediction method that was used to evaluate whether the remaining fault-free thrusters can track the attitude control response performance under the failure of the thruster valve.

The attitude control response includes the following parts:

$$\text{Overshoot} \leq \sigma_*\%$$

$$\text{Rise time} \leq t_{r^*}$$

$$\text{Settling time} \leq t_{s^*}$$

$$\text{Steady-state error} \leq e_*.$$

In the RLV's longitudinal attitude loop, it has a plurality of thrusters for pitch control; the force generated by a single thruster is F, the maximum thrust generated by the thrusters is F_{max}, and the fault boundary of the thruster is F_{hold}, as shown in Figure 5.

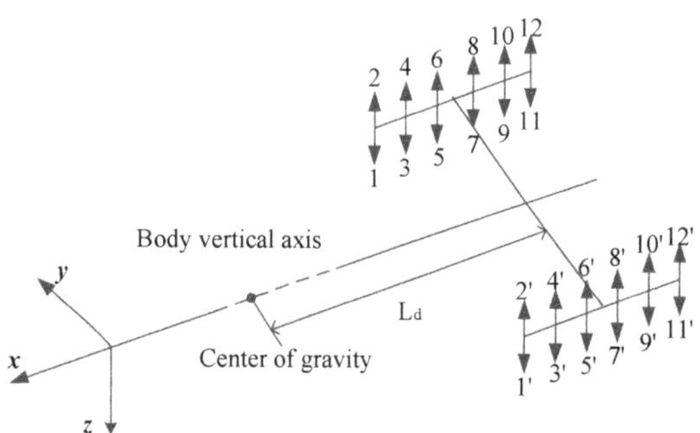

Figure 5. Thruster positions diagrams.

For RLV longitudinal attitude control, the thrusters provide directional impulses in the $\pm y$-axis.

Considering the RLV's longitudinal attitude nonlinear model from Equation (1), M represents the operating pitch moment of thrusters,

$$M = \sum_{i=1}^{m} F_i L_d U_i, \tag{16}$$

where F_i corresponds to the output of the ith thruster, see Equation (2), U_i is the switch command of the ith thruster valve for two states 0 and 1. The command to turn off the thruster valve is 0, and that to turn on the thruster valve is 1. L_d is the distance from the acting force to the center of gravity of the RLV.

In this study, when the fault was detected, the working thrusters were isolated and the other thrusters were employed to track the attitude control response performance. It is important to note that multiple thruster control allocation schemes may satisfy the attitude control response performance; however, minimum consumption of energy is the foremost objective for control allocation strategy:

$$J = \min\{S_1, S_2, \cdots S_n\},$$
$$S_j = \sum_{i=1}^{k} |v_i|, j = 1, 2 \cdots n \qquad (17)$$

where S_j represents each thruster control allocation scheme of energy consumption in a selected time frame, v_i is the number of working thrusters at each control period and J is the priority objective for the control allocation strategy.

An allocation scheme for thruster control that can satisfy the attitude control response of performance was determined in three steps.

The first step was to calculate the gain K_p according to the required rise time after the attitude angle control command was given. The specific algorithm is as follows:

Algorithm 1: K_p gain numerical calculation for rise time requirement

---

Input: α_{cmd}
Output: $\dot{q}$, q, α
Control variable: M
1   *Initialization Equation (1) state and parameters.*
2   **for** $(K_p = n_1$ to $n_2)$ **do**
3   **for** $(t = 0$ to $t_{r^*})$ **do**
4   *RK45 output iterative assignment*
5   **if** $0 \le t \le t_{r^*}$ **then**
6
$$q_{cmd} = (\alpha_{cmd} - \alpha) / (t_{r^*} - t)$$
$$e_q = q_{cmd} - q$$
$$\dot{q}_{cmd} = e_q / K_p$$
$$M_{cmd} = (\dot{q}_{cmd} - \dot{q}) \cdot I_y$$
7   *Determining the thruster control number based on* M_{cmd}
8   **end**
9   *Calculate operating moment input M to RK45*
10   **end**
11   K_p *is determined according to the principle of minimum* $\alpha_{cmd} - \alpha$ *deviation when*
      t_{r^*} *arrives*
12   **end**

---

where q_{cmd} is the pitch rate reference command, α_{cmd} is then given command of the AoA, e_q is the deviation between the pitch rate reference command and the feedback, K_p is the adjustment gain in the limit range $[n_1\ n_2]$ of the RLV's pitch acceleration, $\dot{q}_{cmd}$ is the reference pitch acceleration, and M_{cmd} is the expected control moment. RK45 is the Runge–Kutta method including Equation (1).

The second step is to calculate the control moment and identify the thruster control allocation scheme satisfying the steady-state error requirement based on Algorithm 1. The specific algorithm is as follows:

Algorithm 2: Nonlinear model-based numerical control

Input: α_{cmd}

Output: α

Control variable: M

1   *Initialization Equation (1) state and parameters.*

2   *Calculate K_p by Algorithm 1*

3   **for** (t = 1 **to** N) **do**

4   *RK45 output iterative assignment*

5   **if** $0 \le t \le t_r$ **then**

6   *Calculate operating moment input M to RK45 based on Algorithm 1*

7   **else**

8        $e_\alpha = \alpha_{cmd} - \alpha$

9   **if** $e_\alpha > e_*$ or $e_\alpha < -e_*$ **then**

10   **for** ($j = -n$ **to** n) **do**

11            $Thrst_{num} = j$

12            $\hat{M}_1 = Thrst_{num} \cdot M_0$

13            $e'_j = \alpha_{cmd} - RK45(\hat{M}_1)$

14   **if** $-e_* < e'_j < e_*$ **then**

15   *Save $\hat{M}_1$ to command set*

16   **end**

17   **end**

18   *In the command set, determine $\hat{M}$ by the principle of the minimum energy consumption*

19   **else**

         $\hat{M} = 0$

20   **end**

21   *Calculate operating moment input M to RK45*

22   **end**

23   **end**

---

where e_α is the deviation between the AoA command and the feedback, $Thrst_{num}$ is the remaining fault-free thruster working number, M_0 is the control moment from a single thruster, $\hat{M}$ is the control moment for each control period, and $\hat{M}_1$ is an assumed control moment. The condition of $|e_\alpha| \le e_*$, the algorithm used a circular prediction calculation by the Runge-Kutta method (RK45) to match it with $\hat{M}_1$ input and the flight dynamics model; $\hat{M}_1$ was replaced in turn from rest fault-free thruster $[-n\ n]$ at each cycle to find the optimized control moment.

The third step is to evaluate whether the remaining fault-free thruster that can track the attitude control response performance under the failure of the thruster valve using the nonlinear model-based numerical control prediction. The prediction flow chart is as following Figure 6:

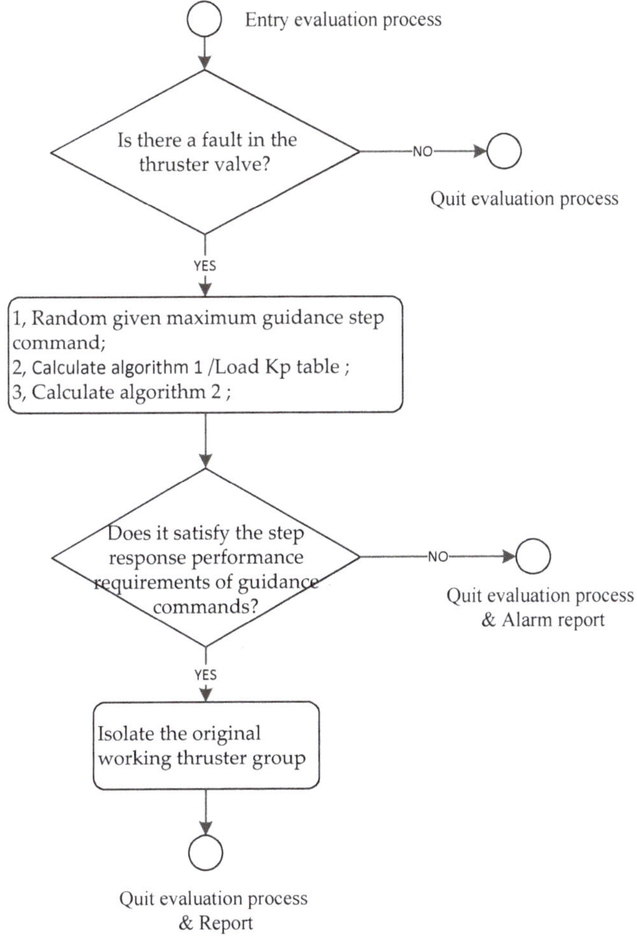

Figure 6. Thruster control recovery evaluation based on fault detection.

In conclusion, the multi-thruster, from a working equivalence point of view, can be regarded as a single thruster. The upper limit of the thruster can be set, and fault detection and isolation can be conducted based on the three types of faults, with the isolation state either turned on or off for all of the types. A numerical calculation method based on a nonlinear model was used to evaluate control recovery and determine if the fault-free thruster could track the attitude control response performance following failure of a thruster valve.

4. Results

In this study, the RLV flight numerical simulation from the deorbit to the initial reentry phase was used to verify the feasibility and performance of the proposed method. The state of the vehicle in the example was as follows: the Mach number $M_a = 25$, AoA initial state was $0°$, AoA command was $\alpha = 40°$, thrust generated by a single thruster $F_i = 3.87$ kN [12], distance from the moment vector to the center of gravity $L_d = 26.86$ m, and moment of inertia $I_v = 651{,}470$ kg·m^2. The nonlinear flight dynamics model in Equation (1) was employed for verification. The 1796 U.S. standard atmosphere model [24] was used in this study and the example vehicle data were comparable to the space shuttle [25].

4.1. Thruster Fault Detection

➢ State estimation

Considering the three fault types A, B, and C, the state estimation and fault-free status comparison were made among the three fault types using the blended Kalman filter, as shown in Figures 7–9.

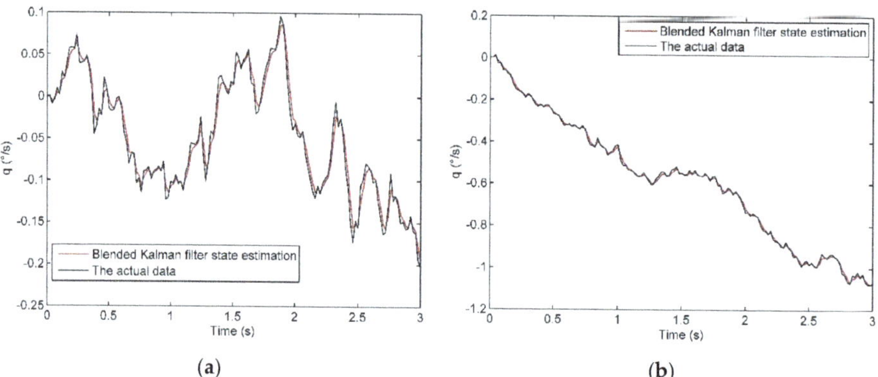

Figure 7. Type A fault vs. fault-free state estimation: (**a**) type A fault state estimation; (**b**) type A fault-free state estimation.

Figure 7 depicts the state estimation of the type A fault and the fault-free state using the blended Kalman filter. The thruster valve was closed in the previous time period, with an initial pitch rate of 0°/s, and, after a command to turn on the valve was given, the fault-free thruster generated a thrust to create an incremental pitch rate, as shown in Figure 7b. Figure 7a displays the type A fault state, that is, it was turned off in the previous moment, with a pitch rate of 0°/s; no response output occurred after a thrust command was given at the latter point as the valve was stuck. As shown in the above figures, the state estimation of the blended Kalman filter was consistent with the actual data trend in the fault and fault-free conditions, and the deviation was in the range of ±0.1°/s.

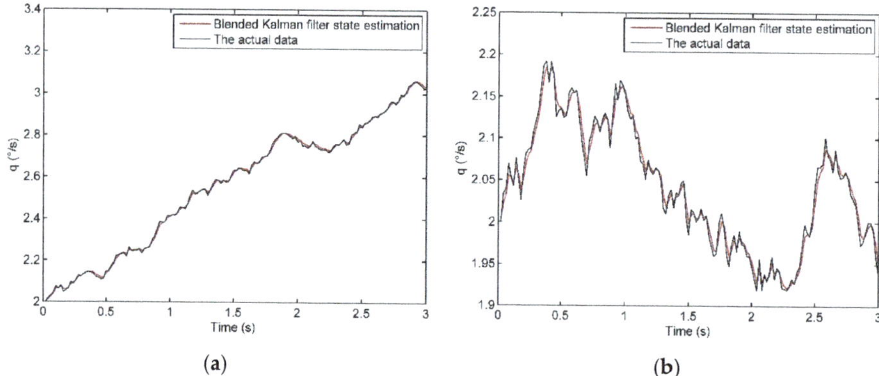

Figure 8. Type B fault vs. fault-free state estimation: (**a**) type B fault state estimation; (**b**) type B fault-free state estimation.

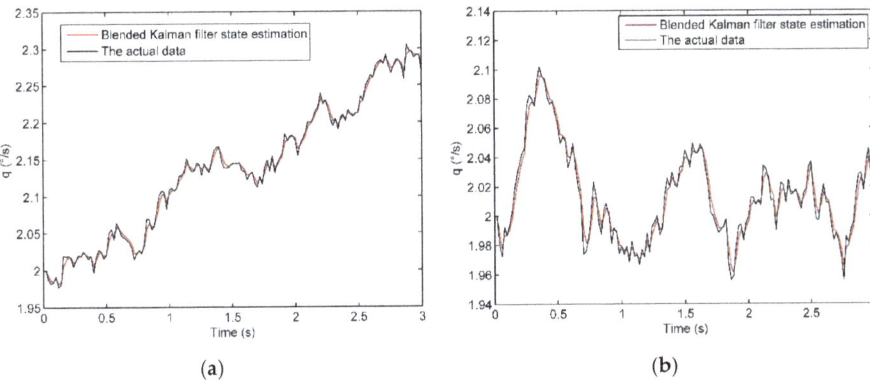

Figure 9. Type C fault vs. fault-free state estimation: (**a**) type C fault state estimation; (**b**) type C fault-free state estimation.

Figure 8 depicts the estimations of the type B fault and the fault-free state. On the premise that the RLV had a pitch rate at the initial moment of 2°/s, the thruster valve in the previous time period was turned on, and, after a command to turn off the valve was given, the thruster should close under the fault-free state and the pitch rate should have been maintained at around 2°/s. Figure 8a displays the type B fault state, that is, after a command to turn off the valve was given, the valve was stuck and did not close. Under the condition of the above-mentioned type B fault and the fault-free state, the state estimation of the blended Kalman filter agreed with the trend of the feedback data, with a deviation within the range of ±0.1°/s.

Figure 9 displays the estimation of the type C fault and the fault-free state. As shown above, when the vehicle had an initial pitch rate of 2°/s, the thruster valve at the previous time was turned on, and, after a command to turn the valve off was given, the thruster did not completely close, indicating a type C fault. Here, it was assumed that the ratio of the valves' incomplete closing distance to the nozzle width was $\tau = 0.2$ after the performance had been attenuated under the limit conditions. As the type C fault is a small thrust-related leakage problem, the pitch rate increment will be slowly generated, as shown in Figure 9a. The pitch rate should fluctuate around the initial pitch rate of 2°/s under the fault-free state, as shown in Figure 9b. The state estimation method adopted in this paper, as mentioned above, remained consistent with the trend of the observed data, with a deviation within the range of ±0.1°/s.

➢   Variance test

In this study, based on the above-mentioned state estimation of the blended Kalman filter, the residual deviation between the state estimation of the pitch rate and the sensor feedback in each period was compared and considered as the fault detection condition. The variance in the 3 s time window was counted; a fault could be ascertained when it was outside of the given threshold. Figure 10 shows a comparison of the residual deviation between the type A, B, and C faults and the fault-free state, respectively, and its variance in the 3 s time window was counted. As shown in Table 1, the minimum variance of type A, B, and C faults was more than 0.01, however the maximum variance of the corresponding fault-free state was less than 0.00006, therefore, a threshold which was an order of magnitude larger was selected as the decision condition to obviously distinguish the fault state from the fault-free state. In addition, the mean test data in Table 1 further shows that a significant difference exists between the mean value of fault-free residual and that of fault residual within 3 s.

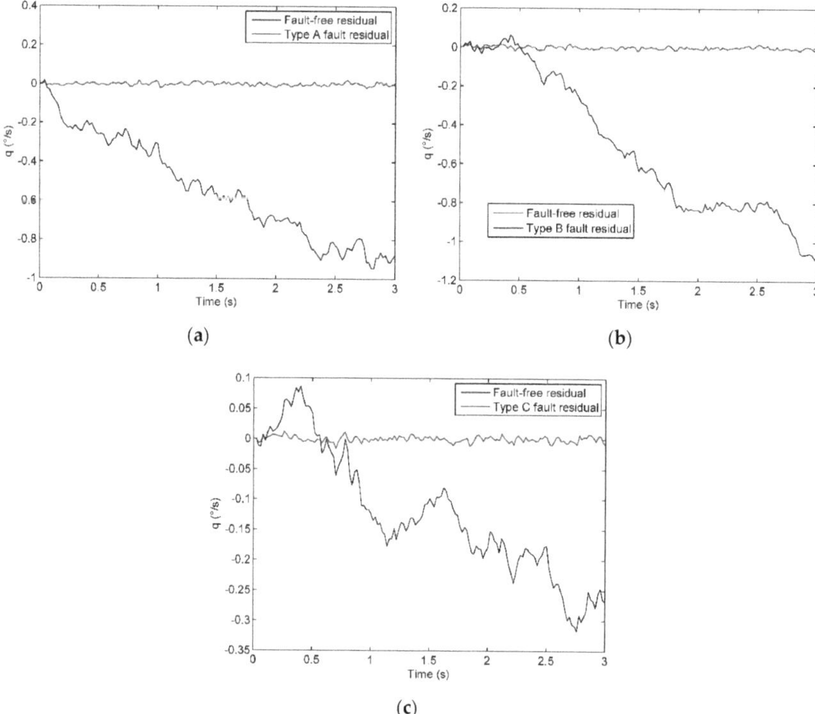

Figure 10. Type A, B, and C fault residuals vs. fault-free residuals: (**a**) type A fault residuals vs. fault-free residuals; (**b**) type B fault residuals vs. fault-free residuals; (**c**) type C fault residuals vs. fault-free residuals.

Table 1. Thruster fault detection.

Fault Type	$D[r_{k\text{-group}}]$	$mean[r_{k\text{-group}}]$	Fault-Free	
			$D[r_{k\text{-group}}]$	$mean[r_{k\text{-group}}]$
A	0.067100	−0.542500	0.000062	−0.000512
B	0.127900	−0.514600	0.000064	−0.000223
C	0.010700	−0.124600	0.000026	−0.000087

➢ False alarm rates and fault detection rates test

In this study, the process noise was zero mean white noise with a standard deviation of $Q^{1/2}$, and the standard deviation of noise changed from 1×10^{-8} to 1×10^{-3}. Each simulation took 1000 data points, chi-square degree of freedom $m = 1$, false alarm rate $\alpha_T = 5\%$, look-up table $T_D = 3.84$. The test results for the false alarm rates are given in Figure 11.

Figure 11a shows that the false alarm rate varies from 3.5 to 6.5, suggesting that the threshold determined by the formula did not meet the requirement ($\alpha_T = 5\%$) as shown in Figure 11a. Increase the correction coefficient β_c, $T_{Dn} = \beta_c T_D$. β_c was obtained through simulation and validation calculations, which increases with 0.01 per step from $\beta_c = 1$. For each β_c test, 10,000 data points were used to verify the false alarm rates $\alpha_T = 5\%$ And from the test, $\beta_c = 1.15$. After correction, Figure 11b shows the false alarm rates varied between 2.5 and 5.0 and met the requirements of the 5% false alarm rate. Due to the existence of approximate control signal noise in the system, the noise is not suppressed after the extended Kalman filter (EKF) based solution is employed. As shown in Figure 11c, the false alarm rates varied from 16 to 26, and these rates are much higher than that of the blending filter solution.

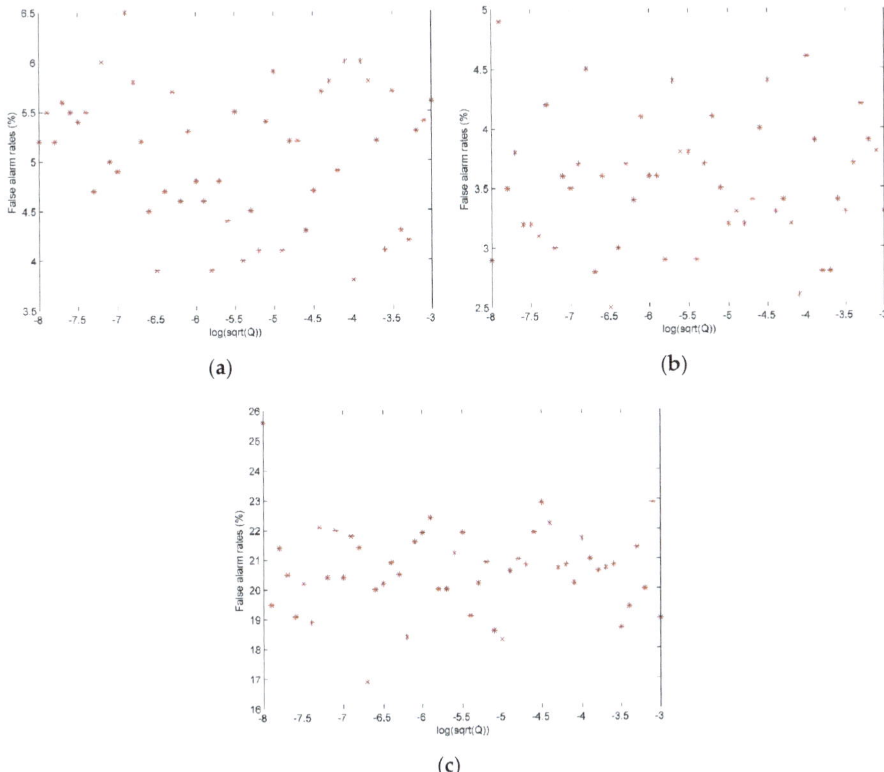

Figure 11. False alarm rates test: (**a**) blending filter solution false alarm rates test by T_D; (**b**) blending filter solution false alarm rates test by correction T_D; (**c**) extended Kalman filter solution false alarm rates test by correction T_D.

Several faults are presented in this paper for the fault detection rates test. The first group was 1000 A-type faults; the second group was 1000 B-type faults; the third group was 1000 C-type faults; the fourth group contained mixed faults i.e., 1000 B-type and C-type mixed faults. The T_{Dn} was used in this test and the test results are given in Table 2.

Based on the classification of class A, B, and C faults, it was easy to distinguish the fault classed based on the state of the control input of class A faults which is different from that of class B and C faults. The class B and C faults produced different thrust so the λ_k from fault detection functions were different and could be classified by comparing their values. All multiple thrusters received the same control command, and the results suggest that class B and C faults can be part of the same group of control thrusters at the same time. Hence, when class B and C failures occur at the same time, the thrust generated is greater making fault detection easier. The blending filter solution has a high detection rate of A-type, B-type, B and C mixed faults i.e., at than 90%, and a good detection rate of 87.8% was obtained for type C leakage faults as they are not easy to detect as shown in Table 2. However, owing to the lack of the suppression function of approximate control signal noise, the EKF based solution showed worse performance than the blending filter solution in the fault detection rates test. The test results are shown in Table 2.

Table 2. Fault detection rates test.

Fault Type	Fault Number	Miss Ratio P_M (%) EKF Solution/ Blending Filter Solution	Fault Detection Rates P_D (%) EKF Solution/ Blending Filter Solution
A	1000	27.40/8.40	72.60/91.60
B	1000	18.00/6.30	82.00/93.70
C	1000	32.10/12.20	67.90/87.80
B&C	1000	10.60/3.80	89.40/96.20

4.2. Thruster Control Recovery Evaluation

This section describes the reference response curve employed to determine whether the remaining fault-free thruster can track the attitude control response performance under the failure of the thruster valve, as shown in Figure 12.

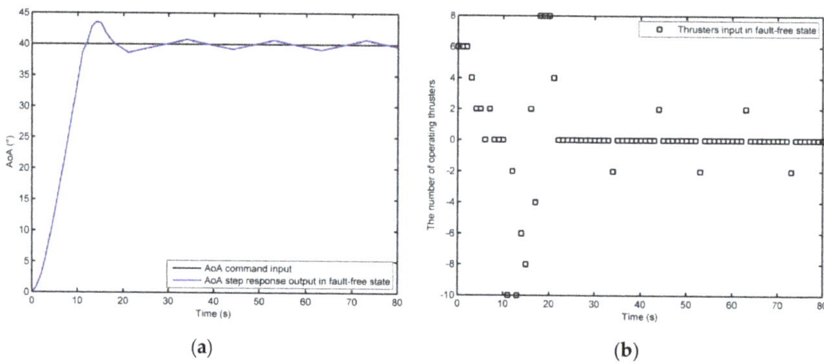

(a) (b)

Figure 12. Angle of attack (AoA) reference response curve and the number of working thrusters in the fault-free state: (**a**) An AoA reference response curve in the fault-free state; (**b**) the number of working thrusters in the fault-free state.

The thruster configuration refers to the space shuttle's data [9]. The RLV includes 38 main thrusters and 6 moving thrusters. The thruster evaluation was studied with reference only to the main thrusters. In accordance with the RLV redundancy configuration, the RLV longitudinal attitude control was configured as $\{-10, \ldots, -2, -1, 0, 1, 2, \ldots, 10\}$, where "$-$" indicates the $+y$-axis impulse generated.

Figure 12 shows an AoA reference response curve and the thruster operation input in the fault-free state. In the rise time period, in order to shorten the rise time and the settling time, more thrusters were needed to generate the expected thrust.

To conduct a simplified analysis of the number of operating thrusters and their operation time, as shown in Figure 12b, 1 s was taken as the unit time, assuming that a thruster was operating at full load for one second, i.e., two states including 1 s continuously turned on or 1 s continuously turned off.

Figure 13 displays the AoA step response output and thruster operation input under type A, B, and C faults and fault-free conditions. During the rise time period, as shown in Figure 13b, the thruster shows a type A fault in the period of 2–5 s and stopped operating. During this period, the blended Kalman filter was used to diagnose and isolate the faulty thruster and, based on the reference response curve, the remaining fault-free thruster was employed to evaluate the control recovery performance using Algorithm 1 and Algorithm 2. The red part in Figure 13b represents the number of thrusters and the working time for the type A fault. During adjustment of the working thruster, the AoA response was as shown in Figure 13a, and to improve the settling time after fault tolerance, both the given pitch rate and the overshoot were increased, while the pitch angle was adjusted by the controller to reach the steady-state requirement.

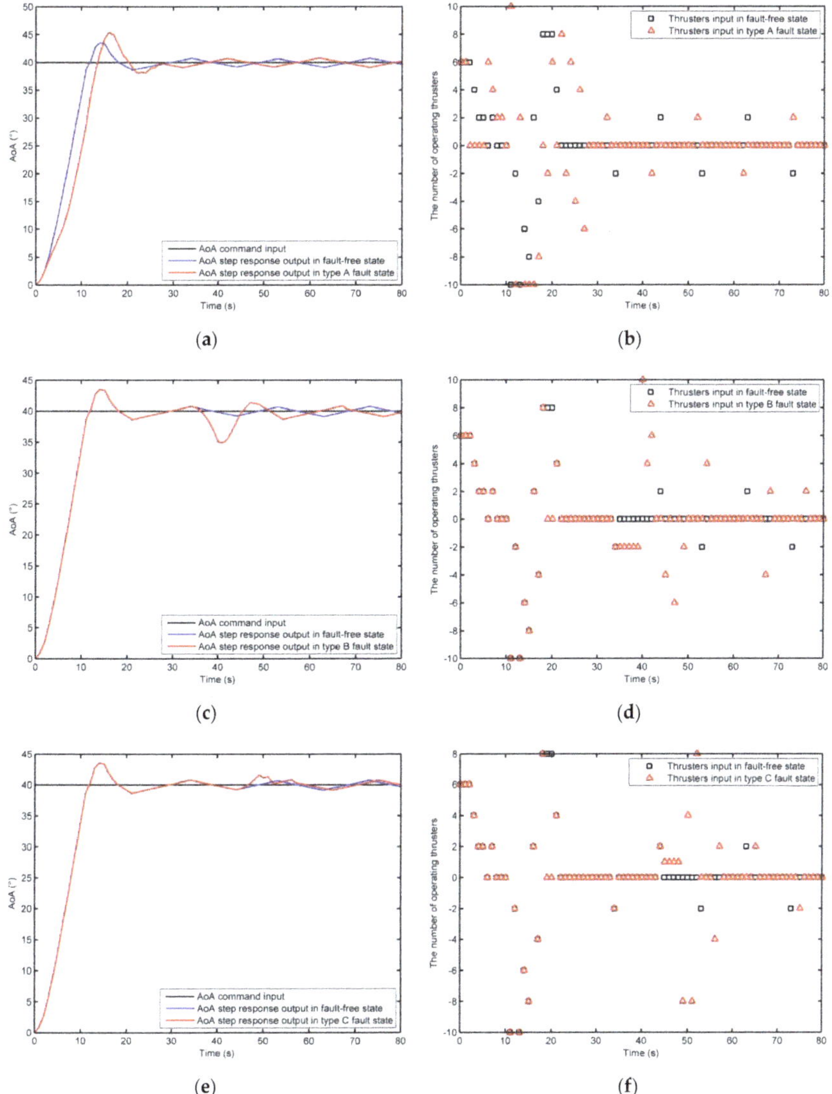

Figure 13. Response output and the number of working thrusters for the type A, B, and C faults and the fault-free state: (**a**) AoA response output for the type A fault and the fault-free state; (**b**) the number of working thrusters in the type A fault and fault-free state; (**c**) AoA response output for the type B fault and the fault-free state; (**d**) the number of working thrusters in the type B fault and fault-free state; (**e**) AoA response output for the type C fault and the fault-free state; (**f**) the number of working thrusters in the type C fault and fault-free state.

In the stage of the attitude angle feedback control, as shown in Figure 13d, the thruster showed a type B fault in the period of 35–39 s, the thruster was turned on at the previous time and did not close at the latter time when the command to turn off was given. The steady-state error then increased, and the controller isolated and switched over the thruster, as shown in Figure 13d. To rapidly reduce the steady-state error, the number of active thrusters was increased, and after the steady-state error was adjusted, the number of thrusters was reduced according to the principle of energy consumption optimization.

In the steady-state stage, as shown in Figure 13f, at 44 s, the two thrusters were in operation, then were closed after a command was given, but in the period of 45–48 s, the thruster was not fully closed due to a type C fault, hence, only one thruster closed. At this stage, the steady-state error increased, and a new demand for operating the thruster was given. After adjustment, the steady-state error stabilized, and the performance requirements were met, as shown in Figure 13e

In the above cases, after the failure of a thruster valve, the remaining fault-free thrusters can satisfy the recovery control performance.

4.3. Running Time and Stability Testing of the Algorithm

This section describes the 1000 cases of random thruster valve failures and AoA command settings that were used to evaluate the running time and stability of the algorithm on the computer, as shown in Figure 14a,b.

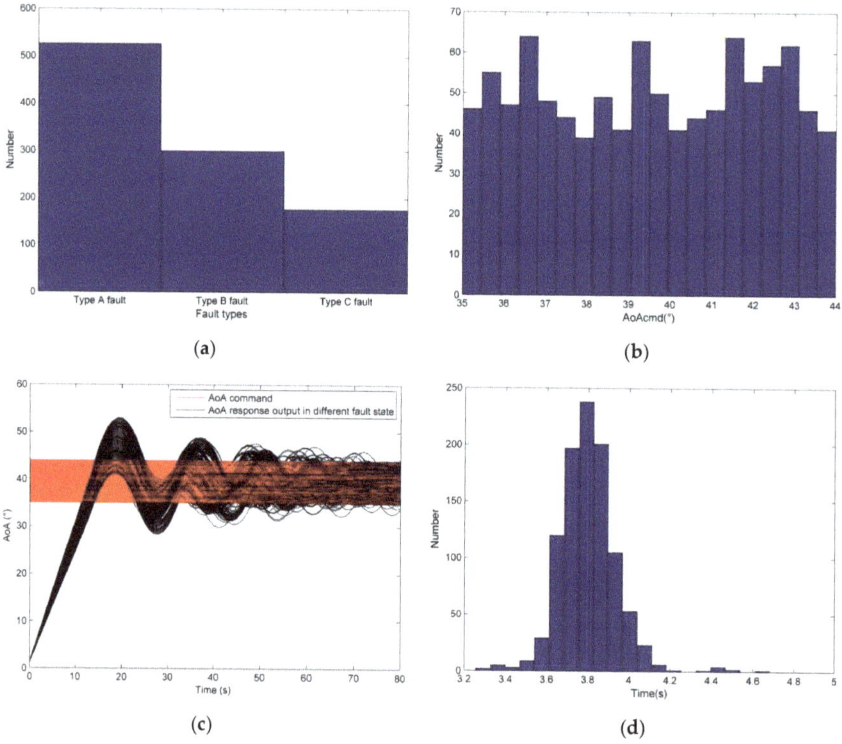

Figure 14. Running time and stability tests of thruster fault detection and control recovery evaluation: (**a**) random fault assignment statistics; (**b**) random AoA command given statistics; (**c**) AoA response output based on the thruster fault detection and control recovery; (**d**) running time statistics for each sample.

The computer configuration is: Intel(R) Core(TM) i5-4200H CPU @ 2.80GHz, RAM 4.00GB, Windows 8 operating system.

Figure 14c shows that the algorithm has good stability in evaluating whether the remaining fault-free thruster can track the attitude control response performance under the failure of a thruster valve. At the same time, Figure 14d shows that the algorithm can complete a fault detection and performance recovery evaluation between 3.2 s and 4.8 s.

5. Discussion

The PWM method is a mature technology of thruster control and is widely used [6–8], but its high-frequency on/off thruster valve operation and valve lifetime are the main causes of the failure problems studied in this paper. Motion-based diagnosis has been proven to be effective [19], and Kalman filtering has been applied in engineering for a long time, however, some shortcomings existed in the proposed research of motion-based fault diagnosis. The approximate control signal noise has a serious impact on Kalman filter model-based state estimation. The frequency of the noise is different from that of the control signal. Therefore, a bandpass filter was introduced to suppress the approximate control signal noise. At the same time, Kalman filtering was used to estimate the motion state and to detect the failure of the thruster valves. The simulation shown in Figure 11 showed that the results met the criterion of the false alarm rates of 5% by blending the filter detection. In addition, to support the redundancy management of the thruster, a recovery evaluation method of employing the remaining fault-free thruster to track the command performance under the failure of a thruster valve was employed; Figure 14 demonstrates the effectiveness of this method. The algorithm running time and stability were tested in 1000 cases, as shown in Figure 14. The running time was between 3.2 s and 4.8 s, and the algorithm operated stably without a dead cycle.

In this study, several thruster missions, including on-orbit, de-orbit, and initial reentry, were assumed to be performed, in which the atmospheric density is close to vacuum. In the future, the extended application of this method, such as the aero-assisted orbital transfer mission, will be focused on, in which the atmospheric density is in the low-density state. In addition, the redundancy management of actuators will also be one of the key research directions in the future.

6. Conclusions

In this paper, the fault types of a thruster valve were analyzed, and three types of thruster valve failure states were examined. By adding a physical filter, the potential influence of the approximate control signal noise on the Kalman filter state estimation was reduced, meanwhile, the residual data was analyzed according to the effective detection and the fault threshold was given in combination with the statistics of the variance. For the three types of fault diagnosis, the threshold generated by the above method was clear and fault detection could be achieved in a short time window. In addition, a controlled recovery evaluation scheme was proposed, combining a nonlinear model-based numerical control prediction, to determine whether the remaining fault-free thrusters can meet the requirements of command performance tracking. Finally, the simulation demonstrated that this method can detect the thruster fault within a certain time window, quickly evaluate the recovery performance, and run stably.

Author Contributions: Conceptualization, H.S.; methodology, H.S.; software, H.S.; validation, H.S. and S.Z.; formal analysis, H.S.; investigation, H.S.; resources, S.Z.; data curation, S.Z.; writing—original draft preparation, H.S.; writing—review and editing, S.Z.; visualization, H.S.; supervision, S.Z.; project administration, S.Z.; funding acquisition, S.Z.

Funding: The paper is supported by the National High Technology Research and Development Program ("863" Program) of China. Tianjin Science and Technology Plan Project under grant number 18YFCZZC00160.

Acknowledgments: We are deeply grateful for the constructive guidance provided by the review experts.

Conflicts of Interest: We declare that there is no conflict of interest regarding the publication of this paper.

References

1. Smith, E.P. Space shuttle in perspective—History in the making. In Proceedings of the AIAA 11th Annual Meeting and Technical Display, Washington, DC, USA, 24–26 February 1975. [CrossRef]

2. Benton, M.; Berry, J.; Benner, H. 2nd generation RLV: Overview of concept development process and results. In Proceedings of the AIAA International Air and Space Symposium and Exposition: The Next 100 Years, Dayton, OH, USA, 14–17 July 2003. [CrossRef]

3. Space.com. Available online: https://www.space.com/25275-x37b-space-plane.html (accessed on 2 June 2017).

4. Jones, S.M.; Reveley, M.S.; Evans, J.K. *Systems Analysis of NASA Aviation Safety Program: Final Report*; NASA/TM-2013-218059; NASA: Hampton, VA, USA, 2013.

5. Astrom, K.J. Where is the intelligence in intelligent control. *IEEE Control Syst.* **1991**, *11*, 37–39.

6. Paradiso, J.A. Adaptable method of managing jets and aerosurfaces for aerospace vehicle control. *J. Guid. Control Dyn.* **1991**, *14*, 44–50. [CrossRef]

7. Thurman, S.W.; Flashner, H. Robust digital autopilot design for spacecraft equipped with pulse-operated thrusters. *J. Guid. Control Dyn.* **1996**, *19*, 1047–1055. [CrossRef]

8. Penchuk, A.N.; Hattis, P.D.; Kubiak, E.T. A frequency domain stability analysis of a phase plane control system. *J. Guid. Control Dyn.* **1985**, *8*, 50–55.

9. Sund, D.C.; Hill, C.S. Reaction control system thrusters for space shuttle orbiters. In Proceedings of the AIAA 15th Joint Propulsion Conference, Las Vegas, NV, USA, 18–20 June 1979; pp. 1144–1150.

10. Zolghadri, A. Advanced model-based FDIR technologies for aerospace systems: Today challenges and opportunities. *Prog. Aerosp. Sci.* **2012**, *53*, 18–29. [CrossRef]

11. Deckert, J.C.; Deyst, J.J. Maximum likelihood failure detection techniques applied to the shuttle orbiter reaction control subsystem. In Proceedings of the AIAA 13th Aerospace Sciences Meeting, Pasadena, CA, USA, 20–22 January 1975; p. 155.

12. Deckert, J.C.; Deyst, J.J. Maximum likelihood failure detection techniques applied to the shuttle RCS jets. *J. Spacecr.* **1976**, *13*, 65–74.

13. Wilson, E.; Rock, S.M. Reconfigurable control of a free-flying space robot using neural networks. In Proceedings of the American Control Conference, Seattle, WA, USA, 21–23 June 1995.

14. Wilson, E. Experiments in Neural Network Control of a Free-Flying Space Robot. Ph.D. Thesis, Mechanical Engineering Department, Stanford University, Stanford, CA, USA, March 1995.

15. Gao, Z.F.; Jiang, B.; Shi, P. Active fault tolerant control design for reusable launch vehicle using adaptive sliding mode technique. *J. Frankl. Inst.* **2012**, *349*, 1543–1560. [CrossRef]

16. Hu, Q.L.; Li, B.; Wang, D.W. Velocity-free fault-tolerant control allocation for flexible spacecraft with redundant thrusters. *Int. J. Syst. Sci.* **2015**, *46*, 967–992. [CrossRef]

17. Zhang, A.H.; Lv, C.C.; Zhang, Z.Q.; She, Z.Y. Finite time fault tolerant attitude control-based observer for a rigid satellite subject to thruster faults. *IEEE Access* **2017**, *5*, 16808–16816. [CrossRef]

18. Liu, C.; George, V.; Sun, Z.W.; Shi, K.K. Observer-based fault-tolerant attitude control for spacecraft with input delay. *J. Guid. Control Dyn.* **2018**, *41*, 2039–2051. [CrossRef]

19. Wilson, E.; Sutter, D.W.; Mah, R.W. Motion-based mass- and thruster-property identification for thruster-controlled spacecraft. In Proceedings of the AIAA Infotech@Aerospace Conference, Arlington, VA, USA, 26–29 September 2005.

20. NASA. *Mitigating in-Space Charging Effects—A Guideline*; NASA-HDBK-4002A; NASA: Hampton, VA, USA, 2011.

21. Cai, G.H.; Song, J.M.; Chen, X.X. Control system design for hypersonic reentry vehicle driven by aerosurfaces and reaction control system. *Proc. Inst. Mech. Eng. Part G J. Aerosp. Eng.* **2015**, *229*, 1575–1587. [CrossRef]

22. Gao, G.Y.; Gao, Y.M.; Dong, Z.H. Preliminary study on visualization of space complex electromagnetic environment. *Mod. Electron. Tech.* **2012**, *35*, 132–139.

23. Chen, Z.; Yang, X.N.; Yang, Y. Overview of GEO satellite under nature strong electromagnetic environment and protection methods. *Environ. Technol.* **2019**, *4*, 68–72.

24. NASA. *U.S. Standard Atmosphere*; U.S. Government Printing Office: Washington, DC, USA, 1976.

25. NASA. Orbiter vehicle. In *Aerodynamic Design Data Book*, 1st ed.; Rockwell International: Milwaukee, WI, USA, 1980; Volume 1L-5.

Article

Fault Diagnosis of Rolling Bearing Using Multiscale Amplitude-Aware Permutation Entropy and Random Forest

Yinsheng Chen [1,2], Tinghao Zhang [3], Wenjie Zhao [1,2], Zhongming Luo [1,2,*] and Kun Sun [1,2]

1   School of Measurement and Communication Engineering, Harbin University of Science and Technology, Harbin 150001, China
2   The Higher Educational Key Laboratory for Measuring & Control Technology and Instrumentations of Heilongjiang Province, Harbin University of Science and Technology, Harbin 150001, China
3   Harbin Institute of Technology, School of Electrical Engineering and Automation, Harbin 150001, China
*   Correspondence: luozhongming@hrbust.edu.cn; Tel.: +86-0451-8639-2308

Received: 9 August 2019; Accepted: 1 September 2019; Published: 4 September 2019

Abstract: A rolling bearing is an important connecting part between rotating machines. It is susceptible to mechanical stress and wear, which affect the running state of bearings. In order to effectively identify the fault types and analyze the fault severity of rolling bearings, a rolling bearing fault diagnosis method based on multiscale amplitude-aware permutation entropy (MAAPE) and random forest is proposed in this paper. The vibration signals of rolling bearings to be analyzed are decomposed into different coarse-grained time series by using the coarse-graining procedure in multiscale entropy, highlighting the fault dynamic characteristics of vibration signals at different scales. The fault features contained in the coarse-grained time series at different time scales are extracted by using amplitude-aware permutation entropy's sensitive characteristics to signal amplitude and frequency changes to form fault feature vectors. The fault feature vector set is used to establish the random forest multi-classifier, and the fault type identification and fault severity analysis of rolling bearings is realized through random forest. In order to demonstrate the feasibility and effectiveness of the proposed method, experiments were fully conducted in this paper. The experimental results show that multiscale amplitude-aware permutation entropy can effectively extract fault features of rolling bearings from vibration signals, and the extracted feature vectors have high separability. Compared with other rolling bearing fault diagnosis methods, the proposed method not only has higher fault type identification accuracy, but also can analyze the fault severity of rolling bearings to some extent. The identification accuracy of four fault types is up to 96.0% and the fault recognition accuracy under different fault severity reached 92.8%.

Keywords: rolling bearings; fault diagnosis; multiscale entropy; amplitude-aware permutation entropy; random forest

---

1. Introduction

A bearing is one of the core components of rotating mechanical system. It is a highly standardized precision mechanical device, which has the advantages of low friction, easy assembly and use, and high work efficiency [1]. However, as a connection between rotating parts, the bearing should also bear a certain load as a supporting part. Therefore, in the use of the bearing, it is inevitably subjected to a series of physical effects such as mechanical stress and mechanical wear, which will lead to deformation and corrosion of the bearing over time, leading to changes in the running state of the bearing. Moreover, due to the imperfect production technology and improper operation and installation, the bearing will also produce artificial damage, and the accumulation and deepening of this damage eventually leads to the failure of bearing [2]. Rolling bearings are mainly composed of inner race, outer race, ball elements and retainer, and the fault types of bearings are generally shown in the form of several parts defects described above, among which the most common types of rolling bearings are the failure of inner race, outer race and ball elements. A bearing is also the most easily worn out part of motor. According to statistics, more than 40% of motor faults are related to a bearing [3,4]. Therefore, fault diagnosis of rolling bearings is of great significance to ensure the reliability and safety of rotating mechanical system.

At present, fault diagnosis methods based on different theories have been successfully applied to fault diagnosis of rolling bearings [5], such as vibration, acoustic noise [6,7], stator current, thermal-imaging, and multiple sensor fusion, etc. The fault diagnosis method based on vibration signal is one of the most commonly used fault diagnosis methods for rolling bearings in rotating machinery. Firstly, it is convenient to acquire vibration signal of a rolling bearing, and the cost of monitoring equipment based on vibrating sensors is low. Secondly, the vibration signal is not easily affected by external environment noise. A rolling bearing inevitably generates vibration along with rotation during operation. Vibration, as an external manifestation of a bearing's own dynamic characteristics, contains rich state information and can reflect bearing fault information. When there is a failure period of the rolling bearing, the vibration signal usually contains some periodic pulse signal. The signal is collected by the vibrating sensor on the bearing base for further analysis and processing, so that the rolling bearing state change information can be obtained and the bearing fault type can be identified. In recent years, scholars in the field of fault diagnosis of rolling bearings based on vibration signal continue to explore, and made a lot of research results. However, there is still the problem of low fault identification accuracy due to insufficient fault feature extraction. Especially, the research of fault severity analysis is one of the urgent problems in this field.

According to the characteristics of vibration signal of rolling bearings, the fault diagnosis methods based on data-driven method, mathematical model and pattern recognition method are widely used [8,9]. Although the data-driven method is flexible in use, it is often used for fault detection and has low fault identification accuracy. The mathematical model-based method needs to model the running state of rolling bearings and accurately model different fault types to obtain better fault diagnosis results. Therefore, pattern recognition is one of the important fault diagnosis methods to analyze the vibration signal of a rolling bearing. It is mainly divided into two parts: effective feature extraction contained in vibration signal and classification of the fault features.

For feature extraction of a rolling bearing vibration signal, the time-frequency analysis has become a principal method to extract the fault characteristics of rolling bearings because the vibration signal is often non-stationary. Wavelet transform (WT) is one of the common time-frequency analysis methods to extract the fault characteristics of rolling bearing vibration signals [10]. The wavelet basis function and kernel parameters need to be predefined, and the different selection of them will influence the effect of feature extraction of vibration signals. Therefore, WT cannot be used to adaptively decompose the signal according to its time-frequency characteristics, which to some extent limits the practical application of WT in rolling bearing fault diagnosis [11]. Wavelet packet transform (WPT) is also a typical time-frequency analysis method for vibration signal of rolling bearings [1]. However, it is necessary to determine the wavelet basis function and the number of decomposition layers of wavelet packet before applying WPT. Improper selection of these will directly affect the effectiveness of wavelet packet decomposition, so that the hidden essential features in the signal cannot be revealed. Empirical mode decomposition (EMD) is a self-adapting time-frequency decomposition technique that was presented by Huang N. E. in 1998 [12]. EMD can adaptively decompose a nonlinear and non-stationary signal into a series of intrinsic mode functions (IMFs) containing signal characteristics of different frequencies. EMD combined with different entropy methods provides an effective method for fault feature extraction of rolling bearings [13–16]. However, EMD is prone to boundary effect and mode mixing problems in the practical application process, which will lead to poor feature extraction effect. In order to avoid the above problems in EMD, ensemble empirical mode decomposition (EEMD), complete ensemble empirical mode decomposition (CEEMD) and local mean decomposition (LMD) have been proposed successively and successfully applied to rolling bearing fault diagnosis [17–22]. Although the above methods in principle solve the boundary effect and mode mixing problem, these problems cannot be completely avoided in real-world application.

The methods based on entropy theory can measure the complexity of complex time series effectively that are often used as the feature extraction methods of rolling bearing vibration signal. The decomposed signals of the above time-frequency analysis methods can be measured by different entropy theories and constitute the feature vector. Complexity of time series can be obtained through several measures such as approximate entropy, sample entropy, energy entropy, permutation entropy and so on. As the calculation result of approximate entropy (ApEn) is excessively dependent on the data length, Richman and Moorman proposed an improved time series complexity measurement method called sample entropy (SampEn) [23]. However, ApEn and SampEn can only measure the complexity of time series from single scale. For many time series, only describing their complexity on a single scale will lose a lot of important information. To more fully describe the complexity of time series, Costa et al. proposed the concept of multiscale entropy (MSE) in 2002 [24]. Firstly, spatial scale segmentation of time series is carried out, and then the SampEn at each scale is calculated, so as to obtain the complexity of time series at different scales. In recent years, combining with the characteristics of other entropy theories, multiscale permutation entropy [25] and multiscale fuzzy entropy [26] have been proposed and successfully applied in the fault feature extraction of rolling bearings. Therefore, it can be seen that the coarse-grained process of MSE to decompose the vibration signal of rolling bearings and using the entropy theory to describe the fault characteristics of the decomposed signals has become an effective way to realize the fault feature extraction of rolling bearings. However, the feature extraction method based on multiscale permutation entropy and multiscale fuzzy entropy ignore the influence of the amplitude of element on entropy in time series, which will make the extracted features have greater randomness and affect the fault identification accuracy. Azami and Escudero proposed the Amplitude-aware permutation entropy (AAPE) to improve the sensitivity of PE to the amplitude and frequency of time series [27]. Different from the permutation entropy algorithm, AAPE algorithm takes into account the mean of signal amplitude and the deviation between amplitudes. In this paper, in order to effectively extract the fault characteristics of rolling bearings, multiscale amplitude-aware permutation entropy (MAAPE) is used to extract the fault features of rolling bearings.

For fault type identification, high performance multi-classifier is used to identify the extracted fault features to determine the fault type or fault severity of rolling bearings. B. Li et al. presented an approach for motor rolling bearing fault diagnosis using neural networks and time-frequency domain bearing vibration analysis [28]. At present, the commonly used artificial neural network includes radical basis function (RBF) neural network and conventional back-propagation (BP) neural network. However, it is difficult to determine the structure of the neural network, since the number of hidden layers is usually selected according to experience. In order to ensure high identification accuracy, a large number of samples are required for training neural network. Support vector machine (SVM) is suitable for solving the classification problem under small sample conditions and has high classification accuracy and good generalization capabilities. Nonetheless, SVM is a kind of binary classifier that needs to use a suitable binary tree to realize multi-classification in the application of multi-classification [11]. Moreover, kernel function selection and kernel parameter setting of SVM have great influence on the identification results. Random forest classifier is an ensemble classifier with high performance that is composed of multiple decision trees [29]. It uses the classification results of all decision trees to determine categories through voting principle. Random forests require very little manual intervention. It does not need to do feature selection or data collation, that is to say, it can determine the features according to the data, so as to simplify the design process of random forest itself. In addition, it usually does not need to preprocess the data in observation data dimension, so it can resist and detect the gross errors that often exist in the form of outlier data. It can maintain a good accuracy when a large proportion of data is lost, and provide the estimation of lost data to simplify its engineering application. Random forests have very fast computing speeds. Since each node of the tree is mainly given by comparison operation, and the amount of calculation is proportional to the depth of a tree, it is very fast to do classification or regression on the well-grown trees.

A fault diagnosis method of rolling bearings based on multiscale amplitude-aware permutation entropy and random forest is proposed in this paper. The vibration of rolling bearings presents the characteristics of periodic instantaneous impact, and the location of bearing failure will lead to different impact characteristics. In order to more fully extract the rolling bearing fault information contained in the vibration signal, using AAPE's sensitivity to the amplitude and frequency changes of vibration signals, the AAPE values at different time scales are calculated as the feature vector, which improves the effectiveness of fault feature extraction. A multiple classifier based on random forests is established by the feature vectors composed of multiscale amplitude-aware permutation entropy to identify the fault type and fault severity of rolling bearings. The contributions of this paper can be summarized as follows:

(1) In this paper, a fault diagnosis method of rolling bearings is presented. On the basis of accurately identifying the fault types of rolling bearings, the fault severity of rolling bearings can be analyzed.
(2) Multiscale amplitude-aware permutation entropy is proposed for the first time, and it is successfully applied to fault feature extraction of rolling bearings.
(3) The random forest multi-classifier is used to identify the fault feature of rolling bearings and analyze the fault severity, and the fault identification accuracy is high.

The rest of this paper is organized as follows. Section 2 introduces the basic principle of the theoretical methods adopted in this paper. The fault diagnosis method of rolling bearings is described in detail in the Section 3. Section 4 introduces the experimental platform and samples, and the experimental results of the proposed fault diagnosis method. Finally, Section 5 summarizes the research content and results of this paper, and prospects the future research content.

2. Methods

2.1. Multiscale Entropy

The MSE is composed of two steps: coarse-graining procedure and computation of SampEn [30].

(1) For the original time series $\{X_i\} = \{x_1, x_2, \cdots, x_n\}$ with length N, coarse granulation is carried out for the given embedded dimension m and similarity tolerance r, so as to form a new coarse-grained time series $y_j^{(\tau)}$,

$$y_j^{(\tau)} = \frac{1}{\tau} \sum_{i=(j-1)\tau+1}^{j\tau} x_i, 1 \leq j \leq N/\tau \tag{1}$$

where $\tau = 1, 2, \cdots, n$ is scale factor. The coarse-graining procedure diagram of MSE is shown in Figure 1.

(2) The sample entropy value of each coarse-grained time series is calculated, and the sample entropy of n coarse-grained time series is expressed as scale factor function as shown below,

$$MSE(x, \tau, m, r) = SE(y_j^\tau, m, r) \tag{2}$$

where similarity tolerance $r = (0.1 \sim 0.25) \cdot Std$ in which Std is the standard deviation of original time series $\{X_i\}$. $SE(y_j^\tau, m, r)$ represents the sample entropy calculation function with scale factor τ, embedded dimension m and similarity tolerance r. The SampEn calculation process can be referred to Ref. [23].

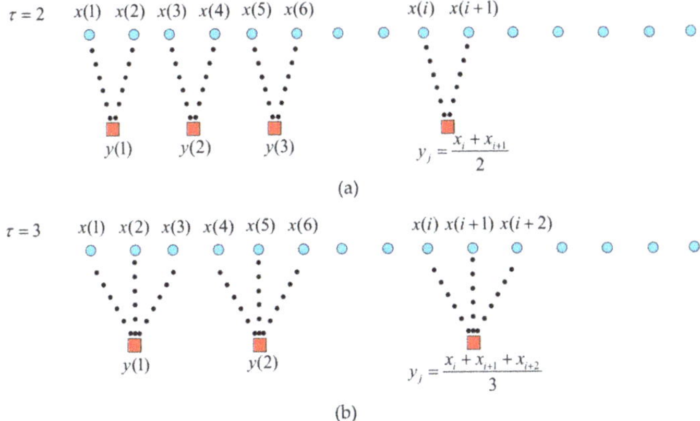

Figure 1. Coarse-graining procedure diagram of multiscale entropy (MSE). (**a**) Scale factor $\tau = 2$; (**b**) Scale factor $\tau = 3$.

2.2. Amplitude-Aware Permutation Entropy

Bandt proposed the concept of Permutation Entropy (PE) in 2002 [31]. Currently PE is widely used in the analysis of complex time series signals to measure the complexity of a nonlinear and non-stationary signal.

Suppose a given time series $x = \{x_1, x_2, \ldots, x_N\}$, and for each time point t, embed signal x into d-dimensional space to obtain reconstructed vector $X_t^{d,l} = \{x_t, x_{t+l}, \ldots, x_{t+(d-2)l}, x_{t+(d-1)l}\}$, $t = 1, 2, \ldots, N - (d - 1)l$ where d and l represent embedded dimension and time delay, respectively. The magnitude of elements in each vector $X_t^{d,l}$ is arranged in ascending order, namely $\{x_{t+(j_1-1)l}, x_{t+(j_2-1)l}, \ldots, x_{t+(j_{d-1}-2)l}, x_{t+(j_d-1)l}\}$, where j_* represents the order of elements in the reconstructed vector $X_t^{d,l}$. Therefore, when the embedding dimension is d, there are a total of $d!$ permutations, and the ith permutation order is demoted as π_i.

The occurrence probability of π_i is expressed as:

$$p(\pi_i) = \frac{f(\pi_i)}{N - d + 1} \tag{3}$$

where $f(\pi_i)$ is a function of counting the number of occurrences of the permutation order π_i. Whenever the permutation order of the elements inside $X_t^{d,l}$ is π_i, $f(\pi_i)$ increases by 1. The definition of permutation entropy is as follows,

$$PE(x, d, l) = -\sum_{\pi_i=1}^{\pi_i=d!} p(\pi_i) \ln p(\pi_i) \tag{4}$$

However, there are two main problems in describing complex time series by permutation entropy. First, traditional PE only considers the order of time series amplitude, but ignores the amplitude information of the corresponding elements in the time series. Second, the effect of elements with equal amplitude on PE value in time series is not clearly explained.

In view of this, literature [27] proposed AAPE to increase PE's sensitivity to amplitude and frequency of time series. Different from the PE algorithm, AAPE algorithm considers the mean value of signal amplitudes and the deviation between amplitudes, and introduces the counting rule of replacing $f(\pi_i)$ in PE with relatively normalized probability.

Assuming that the initial value of $p(\pi_i^{d,l})$ is 0, for the time series $X_t^{d,l}$ in the process of t increasing from 1 to $N - d + 1$, whenever the permutation order is $\pi_i^{d,l}$, $p(\pi_i^{d,l})$ has to be updated.

$$p(\pi_i^{d,l}) = p(\pi_i^{d,l}) + \left(\frac{A}{d} \sum_{k=1}^{d} |x_{t+(k-1)l}| + \frac{1-A}{d-1} \sum_{k=2}^{d} |x_{t+(k-1)l} - x_{t+(k-2)l}| \right) \tag{5}$$

where, $A \in [0, 1]$ is the adjustment coefficient, and the weight of signal amplitude mean and amplitude deviation is adjusted, generally 0.5. So, $p(\pi_i^{d,l})$ is the probability of $\pi_i^{d,l}$ appearing in the whole time series.

$$p(\pi_i^{d,l}) = \frac{p(\pi_i^{d,l})}{\sum_{t=1}^{N-d+1} \left(\frac{A}{d} \sum_{k=1}^{d} |x_{t+(k-1)l}| + \frac{1-A}{d-1} \sum_{k=2}^{d} |x_{t+(k-1)l} - x_{t+(k-2)l}| \right)} \tag{6}$$

The AAPE value of time series is expressed as

$$AAPE(d,l,n) = -\sum_{\pi_k=1}^{\pi_k=d!} p(\pi_k) \ln p(\pi_k) \tag{7}$$

2.3. Random Forest

Random forest (RF) was proposed in 2001 by Breiman, an academician of the national academy of sciences. This algorithm is suitable for solving prediction and classification problems. RF integrates multiple weak classifiers and consists of many decision trees. Its output results are determined by voting principle according to the predicted results of each decision tree in the forest. The basic principle of RF is as follows:

Assuming that the RF classifier is composed of multiple decision trees $\{h_j(x, \Theta_k), k = 1, 2, \ldots, n\}$, $\{\Theta_k, k = 1, 2, \ldots, n\}$ stands for random vectors that are independent and identically distributed. The training sample set of the random forest classifier is expressed as $D = \{(x_1, y_1), (x_2, y_2), \ldots, (x_N, y_N)\}$, where $x_i = (x_{i,1}, \ldots, x_{i,p})^T$ represents that the ith training sample x_i has p eigenvalues, and y_i represents the corresponding label of training sample x_i. The training sample set D is sampled for n times with Bootstrap, and n Bootstrap subsample $D_j(j = 1, 2, \cdots, n)$. For each subsample D_j, decision tree model $h_j(x)$ (CART decision tree is generally used) is constructed, and finally a decision tree classifier composed of a group of decision trees $\{h_1(x), h_2(x), \cdots, h_k(x)\}$ is obtained. For a new test sample, the category that gets the most votes through n decision trees is taken as the final category of the test sample. The classification decision is as follows:

$$f(x) = \text{argmax}_y \sum_{j=1}^{n} I(h_j(x) = y) \tag{8}$$

where $h_j(x)$ represents the jth decision tree, $I(\cdot)$ is indicative function, that is, the value is 1 when this number is in the set, otherwise the value is 0, and y represents the target variable formed by category label y_i.

3. The Proposed Fault Diagnosis Method for Rolling Bearings

This paper presents a fault diagnosis method for rolling bearings based on multiscale amplitude-aware permutation entropy and random forest. Multiscale amplitude-aware permutation entropy is able to extract the fault features of rolling bearing vibration signal at different time scales. Random forest is established by the different fault features to identify the fault type of rolling bearings and analyze its fault severity. The proposed fault diagnosis method for rolling bearings consists of two main processes: training process and testing process. The principle diagram of proposed fault diagnosis method for rolling bearings is shown in Figure 2.

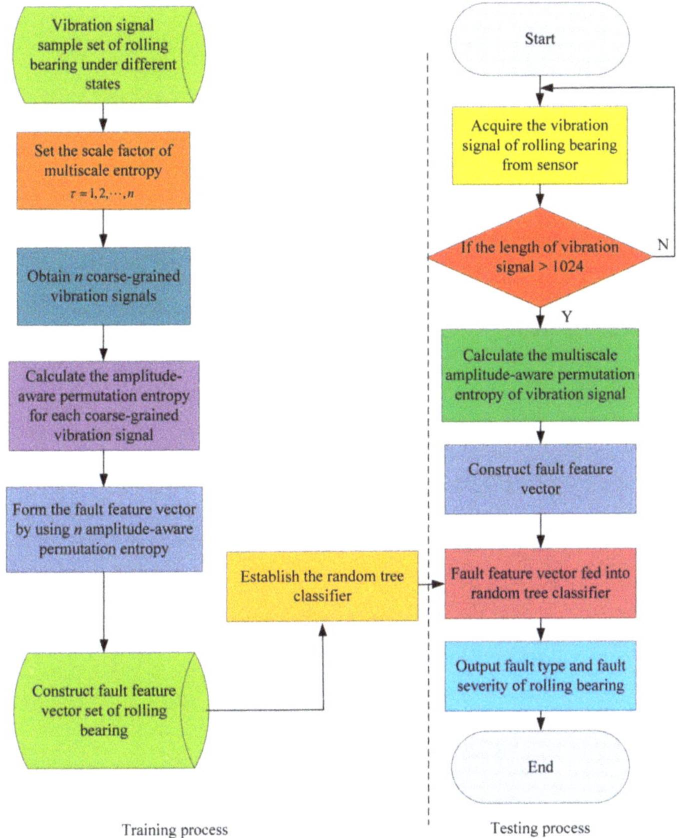

Figure 2. The principle diagram of proposed fault diagnosis method for rolling bearings.

In the training process, the vibration signal sample set of rolling bearings under different states is used to extract the fault feature vectors that can establish the random forest classifier to identify the fault type of rolling bearings. In the testing process, the vibration signal of rolling bearings can be analyzed by the proposed fault diagnosis method, and the fault type and fault severity can be obtained. The main steps of proposed fault diagnosis method are as follows:

In order to ensure the effectiveness of fault diagnosis methods, sufficient samples of rolling bearing vibration signals under different fault states should be collected. These vibration signals construct the vibration signal set.

(1) Set the scale factor $\tau = 1, 2, \cdots, n$ of rolling bearings, and n coarse-grained vibration signals are obtained for each vibration signal in the set.

(2) For each coarse-grained vibration signal, calculate n amplitude-aware permutation entropy values to construct the fault feature vector.

(3) The fault feature vector set is constructed by the fault feature vectors extracted by Step (3).

(4) The random tree classifier is established by the fault feature vector set.

(5) The testing vibration signal can be analyzed by the proposed method and get the fault type and fault severity of rolling bearings.

4. Experiments and Results

4.1. Experimental Setup

In this paper, the rolling bearing fault data set provided by the Case Western Reserve University Bearing Data Center is used to verify the proposed fault diagnosis method [32–34]. In the experiment, Svenska Kullager-Fabriken (SKF) rolling bearing was taken as the research object. The data set collected the rolling bearing vibration signals in four states: normal (NM), inner race (IR) fault, outer race (OR) fault and ball elements (BE) fault through the acceleration sensor. The sampling frequency was 12 kHz for the three types of faults, with diameters of 7 mils, 14 mils and 21 mils that were selected for data acquisition.

The time-domain waveforms of rolling bearing vibration signals with different types and fault severity under 0 load are shown in Figure 3. It can be seen that changes in fault types and fault severity are correlated with changes in amplitude and frequency of vibration signals. Table 1 shows the composition of fault types and fault severity in the experimental sample set, which includes 10 different rolling bearing health states. Each rolling bearing vibration signal is divided into multiple data samples without overlapping, and each sample contains 1024 sampling points to form the experimental data set composed of 50 samples in each healthy state. Among them, 10 samples in each healthy state were used as training set and 40 samples as test set.

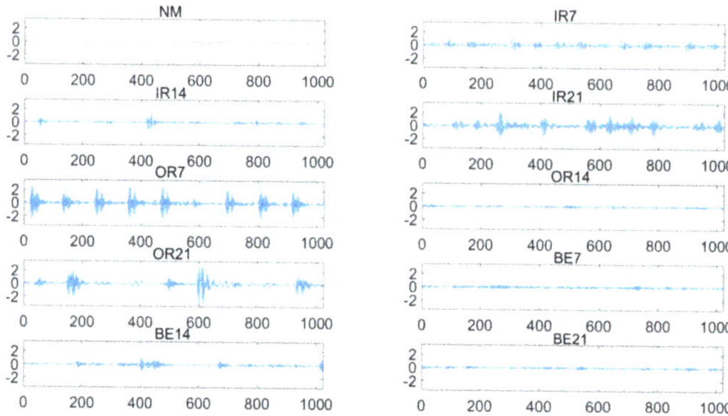

Figure 3. The time-domain waveforms of rolling bearing vibration signals with different types and fault severity under 0 load.

Table 1. The composition of fault types and fault severity in the experimental sample set.

Fault Type	Labels	Fault Diameter (mils)	Load (hp)			
			0	1	2	3
Normal	NM	-	√	√	√	√
Inner race fault	IR07	7	√	√	√	√
	IR14	14	√	√	√	√
	IR21	21	√	√	√	√
Outer race fault	OR07	7	√	√	√	√
	OR14	14	√	√	√	√
	OR21	21	√	√	√	√
Ball elements fault	BE07	7	√	√	√	√
	BE14	14	√	√	√	√
	BE21	21	√	√	√	√

4.2. Experimental Results

The MAAPE-based fault feature extraction method proposed in this paper can extract fault features contained in rolling bearing vibration signals at different scales. In order to highlight the fault characteristics in different time scales of vibration, the scale factor is set as 20 in the experiments of this paper since a too large τ will affect the computation efficiency, while a too small τ cannot extract enough information [26]. The MAAPE feature values of rolling bearing vibration signal under normal condition is shown in Figure 4. From Figure 4, there are obvious differences in the MAAPE feature values at the different scales. AAPE can describe the amplitude and frequency characteristics of vibration signals at different time scales. The characteristics of vibration signals in different time scales are different, which can highlight the fault characteristics more comprehensively.

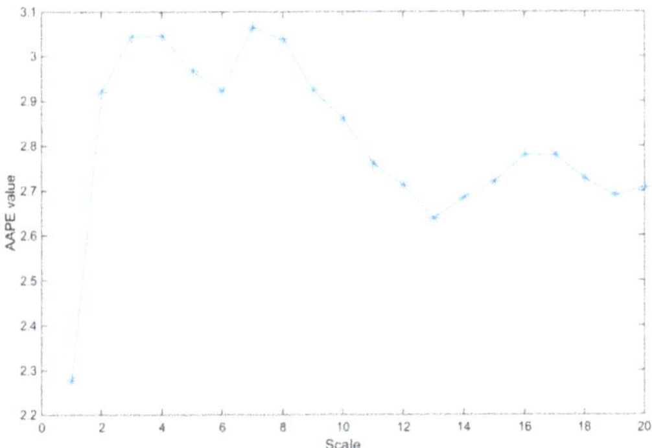

Figure 4. MAAPE feature values of rolling bearing vibration signal under normal condition.

Figure 5 shows the MAAPE feature values of rolling bearing vibration signal under inner race fault condition with different loads. As shown in Figure 5, although the failure size of the inner race is the same and the trend of AAPE values under different scale factors is basically consistent, there are obvious differences in the MAAPE features of rolling bearing vibration signals under different loads. Under different loads, the MAAPE fault feature extraction method can describe the characteristics of vibration signals. In addition, as shown in Figures 4 and 5, it can be seen that in the case of failure, the MAAPE feature extraction method can effectively describe the characteristics of vibration signals under inner race fault condition, and has obvious separability.

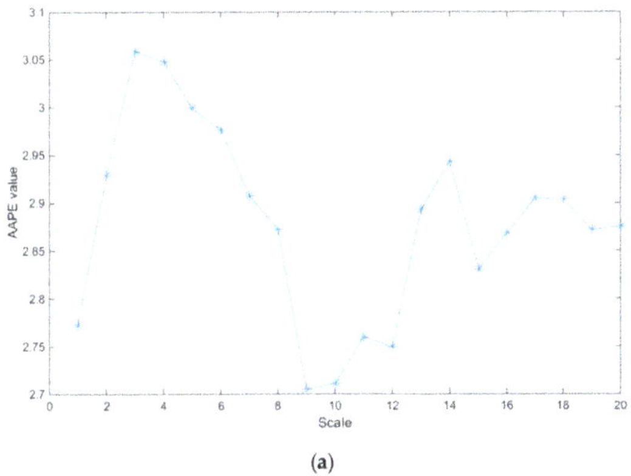

(a)

Figure 5. *Cont.*

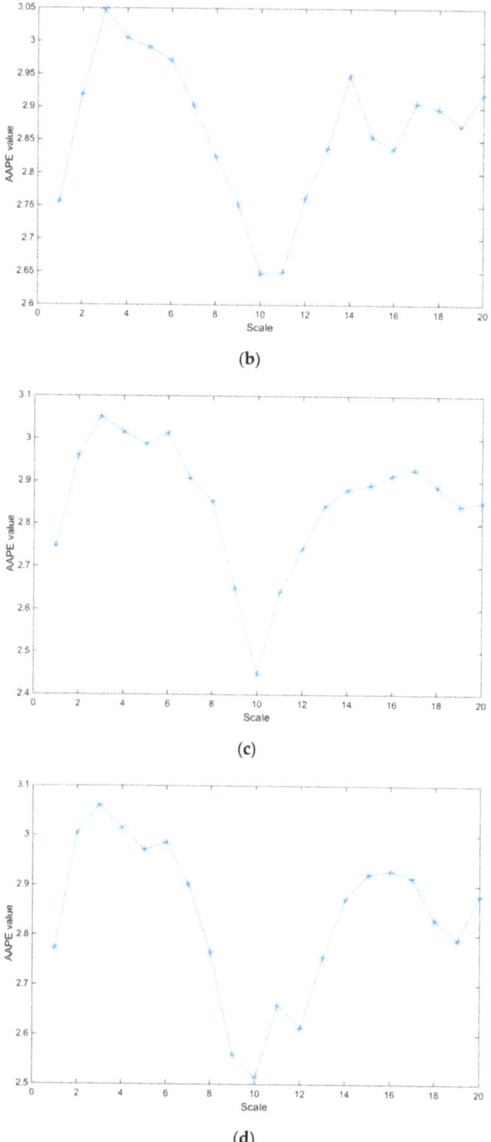

Figure 5. MAAPE feature values of rolling bearing vibration signal under inner race fault condition with different loads. (**a**) Inner race fault, 7 mils, with load 0 hp. (**b**) Inner race fault, 7 mils, with load 1 hp. (**c**) Inner race fault, 7 mils, with load 2 hp. (**d**) Inner race fault, 7 mils, with load 3 hp.

MAAPE feature values of rolling bearing vibration signal under different fault conditions with the same loads are shown in Figure 6. It can be seen that under the same load conditions, although the size of faults is the same, the fault feature vectors extracted by MAAPE have obvious differences. Therefore, MAAPE can effectively extract the feature vector of vibration signals of rolling bearings with different fault types, and the difference of feature vectors is obvious. The fault feature extraction method based on MAAPE proposed in this paper has a good effect on fault feature description.

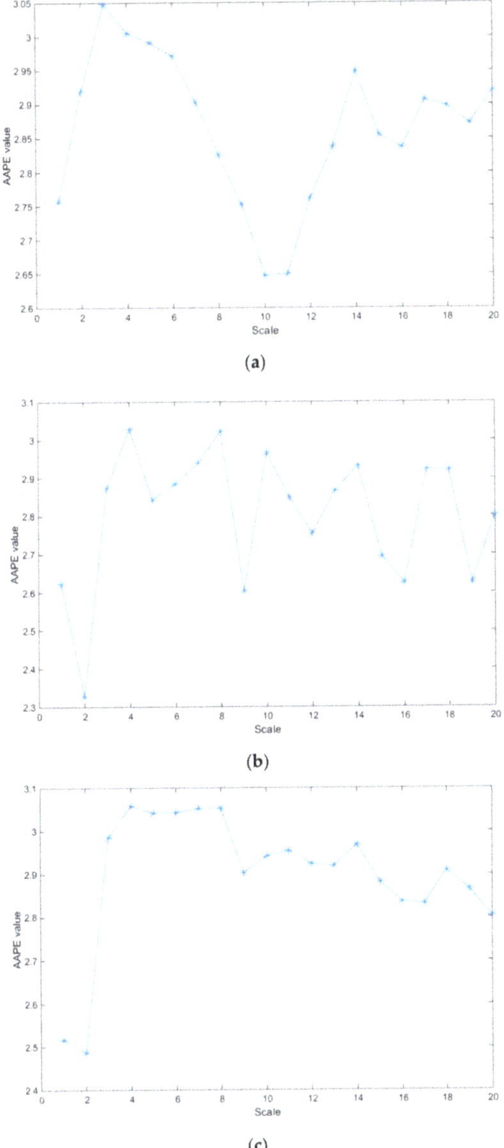

Figure 6. MAAPE feature values of rolling bearing vibration signal under different fault conditions with the same loads. (**a**) Inner race fault, 7 mils, with load 1 hp. (**b**) Outer race fault, 7 mils, with load 1 hp. (**c**) Ball elements fault, 7 mils, with load 1 hp.

Figure 7 shows the feature clustering graph of MAAPE with different loads. In this figure, the 3-dimensional data in the 20-dimensional feature vector constitute the clustering graph. It can be seen that the characteristics of different fault types have certain separability in 3-dimensional space. Therefore, the fault feature extraction method based on MAAPE proposed in this paper can effectively describe the fault features of rolling bearings with different fault types.

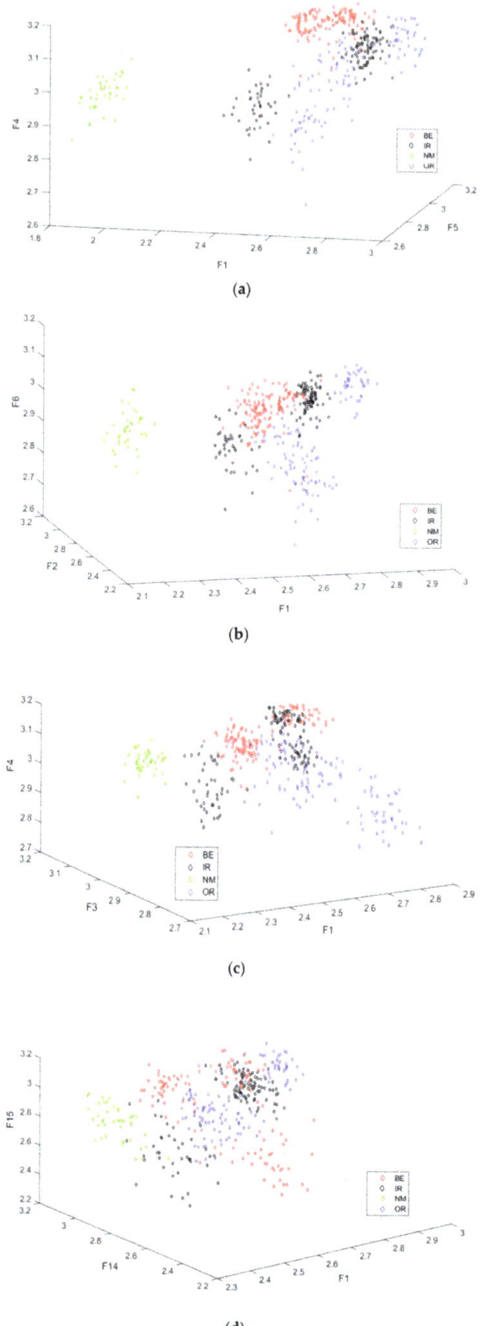

Figure 7. Feature clustering graph of MAAPE with different loads. (**a**) Load 0 hp. (**b**) Load 1 hp. (**c**) Load 2 hp. (**d**) Load 3 hp.

Figure 8 shows that MAAPE feature clustering diagram of each fault severity under different loads. As shown in the figure, the 2-dimensional data in the 20-dimensional feature vector constitute the clustering graph. It can be seen that the fault feature extraction method of MAAPE rolling bearings proposed in this paper has certain feature description effect of fault severity. Fault features extracted under different fault severity have different characteristics which is helpful to identify the fault types and fault severity.

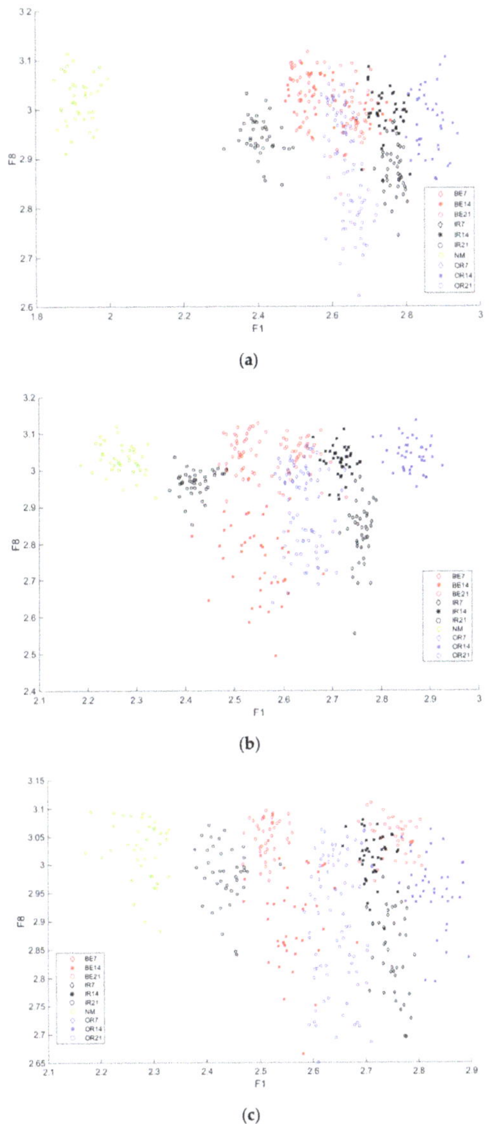

(a)

(b)

(c)

Figure 8. *Cont.*

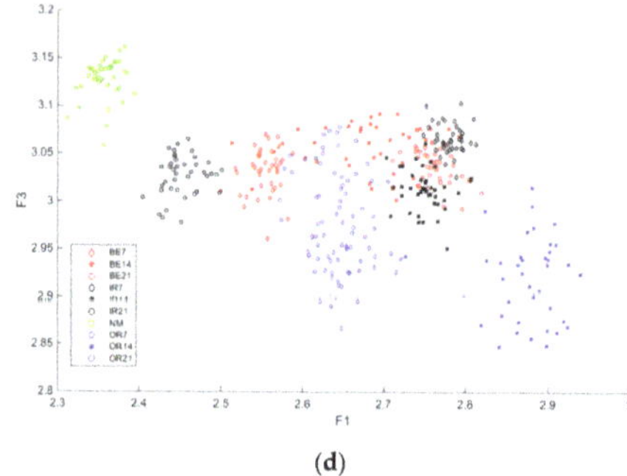

(**d**)

Figure 8. MAAPE feature clustering diagram of each fault severity under different loads. (**a**) Load 0 hp. (**b**) Load 1 hp. (**c**) Load 2 hp. (**d**) Load 3 hp.

Tables 2 and 3 respectively show the experimental results of performance comparison of rolling bearing fault feature extraction algorithms and fault identification accuracy by combining different fault feature extraction methods with random forest. In Table 2, the average between-class distance of MAAPE is less than improved multiscale entropy (IMSE), improved multiscale fuzzy entropy (IMFE) and refined composite multiscale entropy (RCMSE), but the average within-class distance of MAAPE is the smallest. In addition, the average time-consuming of MAAPE is relatively the lowest compared with other feature extraction methods. In Table 3, the fault identification accuracy of MAAPE coupled with random forest is 96.0% and the identification accuracy is slightly lower than RCMPE and IMFE. However, the computational efficiency is the highest as shown in Table 2. Through the above experiments, it can be seen that the rolling bearing fault diagnosis method proposed in this paper has high fault identification accuracy and good execution efficiency, which is more suitable for applications with high real-time requirements.

Table 2. Performance comparison of rolling bearing fault feature extraction algorithms.

Feature Extraction Methods	Average Between-Class Distance	Average Within-Class Distance	Average Time-Consuming (S)
MAAPE	1.24	0.20	0.39
IMPE	0.95	0.23	0.57
RCMPE	0.96	0.30	1.11
IMSE	2.22	0.94	0.42
IMFE	3.72	0.33	17.30
RCMSE	4.02	0.38	0.40

Table 3. Fault identification accuracy by combining different fault feature extraction methods with random forest.

Feature Extraction Methods	Fault Identification Accuracy (%)
MAAPE	96.0%
IMPE	96.0%
RCMPE	97.50%
IMSE	84.25%
IMFE	96.25%
RCMSE	92.25%

In order to illustrate the analytical ability of fault severity of the rolling bearing fault diagnosis method proposed in this paper, the identification accuracy experiments of different fault severity are carried out in this paper, and the experimental results are shown in Table 4. It can be seen from the experimental results that the fault accuracy of the proposed method is low when the ball elements fault severity is 14 mils and 21 mils, and the overall fault accuracy is 92.8%. As the vibration signal contains the natural vibration of rolling bearings, the influence of fault size and external interference, it will have a certain impact on the performance of AAPE fault feature extraction. In addition, the use of 20 different scales of AAPE feature vector to form the eigenvector will increase the uncertainty factor, thereby affecting the accuracy of fault identification. In spite of this, it can be seen from the experimental results that the fault diagnosis method proposed in this paper can analyze the severity of rolling bearing fault to a certain extent.

Table 4. Identification rate of the proposed rolling bearing fault diagnosis method for different fault severity.

Type Labels	NM	IR07	IR14	IR21	OR07	OR14	OR21	BE07	BE14	BE21	Identification Accuracy (%)
NM	40	-	-	-	-	-	-	-	-	-	100
IR07	-	40	-	-	-	-	-	-	-	-	100
IR14	-	-	39	-	-	-	-	-	1	-	97.5
IR21	-	-	-	40	-	-	-	-	-	-	100
OR07	-	-	-	-	40	-	-	-	-	-	100
OR14	-	-	-	-	-	39	-	-	1	-	97.5
OR21	-	1	-	-	-	-	39		-	-	97.5
BE07	-	-	-	-	-	-	-	40	-	-	100
BE14	-	-	-	-	-	2	-	2	34	2	85
BE21	-	-	-	-	-	-	-	4	1	35	87.5

5. Conclusions

A rolling bearing fault diagnosis method based on multiscale amplitude-aware permutation entropy (MAAPE) and random forest is proposed in this paper. The MAAPE-based fault feature extraction method is proposed for the first time. In order to effectively extract fault features in vibration signals of rolling bearings, this paper uses a coarse-grained process in MSE to obtain a set of coarse-grained time series under different scale factors, and extracts fault features contained in different coarse-grained time series through AAPE to obtain feature vectors representing fault information of rolling bearings. Then, the random forest classifier is adopted to identify the fault feature vector extracted from vibration signal of rolling bearings. The proposed fault diagnosis method can effectively identify the fault type of rolling bearings and the identification accuracy of four fault types is up to 96.0%. The proposed method can further analyze the fault severity of rolling bearings, and the identification accuracy under different fault severity is 92.8%.

MAAPE can improve the stability of PE, become more sensitive to amplitude and frequency changes, and contribute to the description of fault characteristics. However, because it is too sensitive to the change of amplitude, it is easy for external interference to affect the extracted features, leading to a decline in identification accuracy. Therefore, how to avoid external interference and highlight bearing vibration signals before feature extraction will help improve the feature extraction effect of MAAPE. The following research direction is to further improve the fault feature description ability of MAAPE, and then improve the fault severity analysis ability of fault diagnosis method.

Author Contributions: Y.C. clarified the research content and ideas, and completed the writing of this paper. T.Z. and W.Z. studied the method and carried out experiments. Z.L. and K.S. analyzed the data and proofread the manuscript.

Funding: This research was funded by the National Natural Science Foundation of China under grant numbers 61803128, 61801148 and Heilongjiang Province Natural Science Foundation under grant numbers F2016021 and LH2019F026.

Acknowledgments: Thanks for the Bearing Data Center of Case Western Reserve University for supplying the rolling bearing data set.

Conflicts of Interest: The authors declare no conflict of interest.

Nomenclature

MAAPE	Multiscale amplitude-aware permutation entropy
WT	Wavelet transform
WPT	Wavelet packet transform
EMD	Empirical mode decomposition
EEMD	Ensemble empirical mode decomposition
CEEMD	Complete ensemble empirical mode decomposition
LMD	Local mean decomposition
ApEn	Approximate entropy
SampEn	Sample entropy
MSE	Multiscale entropy
PE	Permutation entropy
RBF	Radical basis function
BP	Back-propagation
SVM	Support vector machine
SKF	Svenska Kullager-Fabriken
RF	Random forest
IMPE	Improved multiscale permutation entropy
RCMPE	Refined composite multiscale permutation entropy
IMSE	Improved multiscale entropy
IMFE	Improved multiscale fuzzy entropy
RCMSE	Refined composite multiscale entropy

References

1. Chen, Y.; Zhang, T.; Luo, Z.; Sun, K. A novel rolling bearing fault diagnosis and severity analysis method. *Appl. Sci.* **2019**, *9*, 2356. [CrossRef]
2. Shi, J.; Liang, M.; Guan, Y. Bearing fault diagnosis under variable rotational speed via the joint application of windowed fractal dimension transform and generalized demodulation: A method free from prefiltering and resampling. *Mech. Syst. Signal Process.* **2016**, *68*, 15–33. [CrossRef]
3. Georgoulas, G.; Loutas, T.; Stylios, C.D.; Kostopoulos, V. Bearing fault detection based on hybrid ensemble detector and empirical mode decomposition. *Mech. Syst. Signal Process.* **2013**, *41*, 510–525. [CrossRef]
4. Cerrada, M.; Sánchez, R.-V.; Li, C.; Pacheco, F.; Cabrera, D.; de Oliveira, J.V.; Vásquez, R.E. A review on data-driven fault severity assessment in rolling bearings. *Mech. Syst. Signal Process.* **2018**, *99*, 169–196. [CrossRef]
5. Zhang, S.; Zhang, S.; Wang, B.; Habetler, T.G. Machine learning and deep learning algorithms for bearing fault diagnostics—A comprehensive review. *arXiv* **2019**, arXiv:1901.08247. Available online: https://arxiv.org/abs/1901.08247 (accessed on 11 August 2019).
6. Glowacz, A.; Glowacz, W.; Glowacz, Z.; Kozik, J. Early fault diagnosis of bearing and stator faults of the single-phase induction motor using acoustic signals. *Measurement* **2018**, *113*, 1–9. [CrossRef]
7. Glowacz, A. Fault diagnosis of single-phase induction motor based on acoustic signals. *Mech. Syst. Signal Process.* **2019**, *117*, 65–80. [CrossRef]
8. Martínez-García, C.; Astorga-Zaragoza, C.; Puig, V.; Reyes-Reyes, J.; López-Estrada, F. A simple nonlinear observer for state and unknown input estimation: DC motor applications. *IEEE T. Circuits-II.* **2019**, 1.
9. De Moura, E.P.; Souto, C.R.; Silva, A.A.; Irmão, M.A.S. Evaluation of principal component analysis and neural network performance for bearing fault diagnosis from vibration signal processed by RS and DF analyses. *Mech. Syst. Signal Process.* **2011**, *25*, 1765–1772. [CrossRef]
10. Yan, R.; Gao, R.X.; Chen, X. Wavelets for fault diagnosis of rotary machines: A review with applications. *Signal Process.* **2014**, *96*, 1–15. [CrossRef]
11. Li, Y.; Xu, M.; Wei, Y.; Huang, W. A new rolling bearing fault diagnosis method based on multiscale permutation entropy and improved support vector machine based binary tree. *Measurement* **2016**, *77*, 80–94. [CrossRef]

12. Huang, N.E.; Shen, Z.; Long, S.R.; Wu, M.C.; Shih, H.H.; Zheng, Q.; Yen, N.-C.; Tung, C.C.; Liu, H.H. The empirical mode decomposition and the Hilbert spectrum for nonlinear and non-stationary time series analysis. *Proc. Math. Phys. Eng. Sci.* **1998**, *454*, 903–995. [CrossRef]
13. Ali, J.B.; Fnaiech, N.; Saidi, L.; Chebel-Morello, B. Application of empirical mode decomposition and artificial neural network for automatic bearing fault diagnosis based on vibration signals. *Appl. Acoust.* **2015**, *89*, 16–27.
14. Li, Y.X.; Li, Y.A.; Chen, Z.; Chen, X. Feature extraction of Ship-Radiated noise based on permutation entropy of the intrinsic mode function with the highest energy. *Entropy* **2016**, *18*, 393. [CrossRef]
15. Huang, Y.; Wang, K.; Zhou, Q.; Fang, J.; Zhou, Z. Feature extraction for gas metal arc welding based on EMD and time–frequency entropy. *Int. J. Adv. Manuf. Tech.* **2017**, *92*, 1439–1448. [CrossRef]
16. Lei, Y.; Lin, J.; He, Z.; Zuo, M.J. A review on empirical mode decomposition in fault diagnosis of rotating machinery. *Mech. Syst. Signal Process.* **2013**, *35*, 108–126. [CrossRef]
17. Zhang, X.; Liang, Y.; Zhou, J.; Zang, Y. A novel bearing fault diagnosis model integrated permutation entropy, ensemble empirical mode decomposition and optimized SVM. *Measurement* **2015**, *69*, 164–179. [CrossRef]
18. Fu, Q.; Jing, B.; He, P.; Si, S.; Wang, Y. Fault feature selection and diagnosis of rolling bearings based on EEMD and optimized Elman_AdaBoost algorithm. *IEEE Sens. J.* **2018**, *18*, 5024–5034. [CrossRef]
19. Zhou, S.; Qian, S.; Chang, W.; Xiao, Y.; Cheng, Y. A novel bearing multi-fault diagnosis approach based on weighted permutation entropy and an improved SVM ensemble classifier. *Sensors* **2018**, *18*, 1934. [CrossRef]
20. Abdelkader, R.; Kaddour, A.; Bendiabdellah, A.; Derouiche, Z. Rolling bearing fault diagnosis based on an improved denoising method using the complete ensemble empirical mode decomposition and the optimized thresholding operation. *IEEE Sens. J.* **2018**, *18*, 7166–7172. [CrossRef]
21. Han, M.; Pan, J. A fault diagnosis method combined with LMD, sample entropy and energy ratio for roller bearings. *Measurement* **2015**, *76*, 7–19. [CrossRef]
22. Sun, J.; Xiao, Q.; Wen, J.; Wang, F. Natural gas pipeline small leakage feature extraction and recognition based on LMD envelope spectrum entropy and SVM. *Measurement* **2014**, *55*, 434–443. [CrossRef]
23. Richman, J.S.; Moorman, J.R. Physiological time-series analysis using approximate entropy and sample entropy. *Am. J. Physiol. Heart Circ. Physiol.* **2000**, *278*, H2039. [CrossRef]
24. Costa, M.; Goldberger, A.L.; Peng, C.K. Multiscale entropy analysis of complex physiologic time series. *Phys. Rev. Lett.* **2002**, *89*, 068102. [CrossRef]
25. Morabito, F.C.; Labate, D.; Foresta, F.L.; Bramanti, A.; Morabito, G.; Palamara, I. Multivariate multi-scale permutation entropy for complexity analysis of Alzheimer's disease EEG. *Entropy* **2012**, *14*, 1186–1202. [CrossRef]
26. Li, Y.; Xu, M.; Wang, R.; Huang, W. A fault diagnosis scheme for rolling bearing based on local mean decomposition and improved multiscale fuzzy entropy. *J. Sound Vib.* **2016**, *360*, 277–299. [CrossRef]
27. Azami, H.; Escudero, J. Amplitude-aware permutation entropy: Illustration in spike detection and signal segmentation. *Comput. Meth. Programs Biomed.* **2016**, *128*, 40–51. [CrossRef]
28. Li, B.; Chow, M.Y.; Tipsuwan, Y.; Hung, J.C. Neural-network-based motor rolling bearing fault diagnosis. *IEEE Trans. Ind. Electron.* **2002**, *47*, 1060–1069. [CrossRef]
29. Breiman, L. *Random Forests. Machine Learning*; Springer: Berlin, Germany, 2001; pp. 5–32.
30. Humeau-Heurtier, A. The multiscale entropy algorithm and its variants: A review. *Entropy* **2015**, *17*, 3110–3123. [CrossRef]
31. Bandt, C.; Pompe, B. Permutation entropy: A natural complexity measure for time series. *Phys. Rev. Lett.* **2002**, *88*, 174102. [CrossRef]
32. Bearing Data Center, Case Western Reserve University. Available online: http://csegroups.case.edu/bearingdatacenter/pages/download-data-file (accessed on 31 January 2010).
33. Li, Y.; Wang, X.; Si, S.; Huang, S. Entropy based fault classification using the Case Western Reserve University data: A benchmark study. *IEEE Trans. Reliab.* **2019**, 1–14. [CrossRef]
34. Li, Y.; Wang, X.; Liu, Z.; Liang, X.; Si, S. The entropy algorithm and its variants in the fault diagnosis of rotating machinery: A review. *IEEE Access* **2018**, *6*, 66723–66741. [CrossRef]

Article

An Intelligent Warning Method for Diagnosing Underwater Structural Damage

Kexin Li [1,*], Jun Wang [1,*] and Dawei Qi [2,*]

1   College of Civil Engineering, Northeast Forestry University, Harbin 150040, China
2   College of Science, Northeast Forestry University, Harbin 150040, China
*   Correspondence: like@nefu.edu.cn (K.L.); jun.w.619@163.com (J.W.); qidw9806@126.com (D.Q.)

Received: 16 July 2019; Accepted: 26 August 2019; Published: 30 August 2019

Abstract: A number of intelligent warning techniques have been implemented for detecting underwater infrastructure diagnosis to partially replace human-conducted on-site inspections. However, the extensively varying real-world situation (e.g., the adverse environmental conditions, the limited sample space, and the complex defect types) can lead to challenges to the wide adoption of intelligent warning techniques. To overcome these challenges, this paper proposed an intelligent algorithm combing gray level co-occurrence matrix (GLCM) with self-organization map (SOM) for accurate diagnosis of the underwater structural damage. In order to optimize the generative criterion for GLCM construction, a triangle algorithm was proposed based on orthogonal experiments. The constructed GLCM were utilized to evaluate the texture features of the regions of interest (ROI) of micro-injury images of underwater structures and extracted damage image texture characteristic parameters. The digital feature screening (DFS) method was used to obtain the most relevant features as the input for the SOM network. According to the unique topology information of the SOM network, the classification result, recognition efficiency, parameters, such as the network layer number, hidden layer node, and learning step, were optimized. The robustness and adaptability of the proposed approach were tested on underwater structure images through the DFS method. The results showed that the proposed method revealed quite better performances and can diagnose structure damage in underwater realistic situations.

Keywords: structural health monitoring; digital image processing; damage; gray level co-occurrence matrix; self-organization map

1. Introduction

Underwater structures, including bridges and dams, are becoming susceptible to deterioration as they are interfere with irresistible factors, such as wind, wave, corrosion, and hydraulic erosion. This inevitable process signifies urgent warning issues, which have motivated people to warn these structures on a regular basis. Thus, many research groups have proposed structural damage warning techniques [1–3]. Structural damage warning is an important part in the process of damage identification, and it is a great challenge to solve the damage detection problem of large complex engineering structures [4–6].

Recently, few satisfactory schemes for underwater structures damage warning have been proposed [7,8]. One possible solution is combining visual inspection with machine learning [9]. Rutkowski and Prokopiuk proposed the identification of the contamination source location in the water distribution system (WDS) based on the learning vector quantization (LVQ) neural network classifier [10].

Chatterjee proposed a neural network training method based on particle swarm optimization, which can solve the problem of multi-layer reinforced concrete building structure damage prediction by detecting the future damage possibility of a multi-layer reinforced concrete building structure [11]. Kai Xu et al. proposed a new approach to damage detection of a concrete column structure subjected to blast loads using embedded piezoceramic smart aggregates [12]. According to research by Feng et al., almost all types of structural damage can be effectively identified by image processing [13–15]. Zhu Z et al. proposed a new retrieval method for crack defect performance, which located the crack point by the most advanced crack detection technology and used image refinement technology to identify the skeleton structure of the point [16]. Molero et al. applied ultrasound imaging to assess the extent of damage during concrete freeze–thaw cycles and extracted NAAP (damage index) parameters as an evaluation criterion [17]. German et al. presented a new method for automatic detection of the flaking area of reinforced concrete columns and measured its characteristics in image data [18]. Based on the data provided by self-powered wireless sensors, Hasni et al. proposed a steel beam tensional vibration fatigue cracking detection method based on artificial intelligence [19].

Although these methods have achieved superior results through the combination of image properties and neural network models for damage identification, it is difficult to realize accurate damage diagnosis for underwater real-world structures [20–23]. Taking image-labeled data from underwater real-world situations is restricted and the identification accuracy is influenced by environmental conditions and the defect type. The gray level co-occurrence matrix is a typical algorithm in statistical analysis and is widely used in various fields because of its good discrimination ability [24,25]. As an important pattern recognition method, self-organization map neural network (SOM) has been widely used in automatic identification of signals, texts, images, and other fields due to its illuminating memory, learning, and generalization [26–28]. Additionally, the combination of gray level co-occurrence matrix (GLCM) and self-organization map (SOM) is widely used for tumor diagnosis, wood defect classification, and cloud classification [29–32]. However, there are few algorithms for the detection of damage in underwater concrete structures that combine the two methods. Therefore, an intelligent warning method is essential for dealing with the accurate diagnose of structural damage problems. This paper combines GLCM and the SOM neural network to improve the recognition performance of underwater structure damage.

2. Methodology

The construction of GLCM is proposed for characterizing the micro-damage of underwater structures in order to obtain an effective micro-damage diagnostic factor. Figure 1 shows micro-damage patterns of common underwater structures. Different types of micro-damages have a variety of texture primitives and exhibit their respective texture characteristics, obtained according to the analysis of the image properties [33–36]. GLCM can evaluate the overall situation of different damage types in gray space comprehensively. It is necessary to use GLCM as a texture analysis method to characterize the apparent micro-damage morphology of underwater structures.

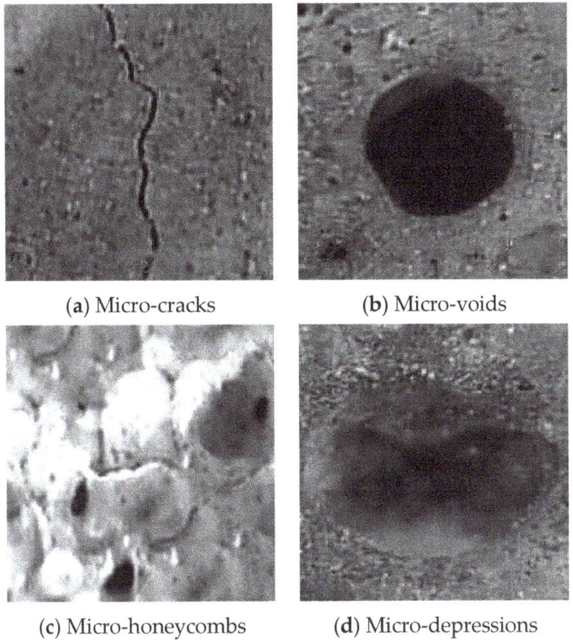

(**a**) Micro-cracks (**b**) Micro-voids

(**c**) Micro-honeycombs (**d**) Micro-depressions

Figure 1. Micro-damage patterns.

2.1. Gray Level Co-Occurrence Matrix

The gray level co-occurrence matrix (GLCM), which specifies a certain condition in the gray space, counts the number of probability occurrences. Additionally, it meets the specified condition in the entire image of the pixel pair. Suppose the pixel pair coordinates are in order (x, y), $(x + dx, y + dy)$ and the pixel values are $f(x, y) = i$, $f(x + dx, y + dy) = j$. When the image gray level is g, the criterion of the sliding window is defined as follows. The pixel pair spacing value is d, and the angle between the connecting line and the horizontal direction is θ. Among them, g, d, and θ are generative criterion of the GLCM. $P(i, j, d, \theta)$ is the probability that pairs of pixels appear at the same time. The GLCM can be expressed by Equation (1):

$$P(i, j, d, \theta) = \left\{ \left[(x, y), (x + dx, y + dy) \middle| f(x, y) = i, f(x + dx, y + dy) = j \right] \right\}, \tag{1}$$

where θ is $0°$, $45°$, $90°$, and $135°$.

Besides, dx, dy are the offset vectors; the equation is as follows:

$$dx = d \cos \theta, \; dy = d \sin \theta. \tag{2}$$

2.2. Feature Parameters

In order to quantitatively describe the direction, trend, and complexity of different micro-damage types in gray space, the gray-scale co-occurrence matrix was applied for feature extraction [30]. Table 1 shows the calculation formula and texture characteristic of the parameters.

Table 1. Formula of parameters and their texture features.

NO.	Parameters	Calculation Formulas	Texture Characteristic
T_1	Angular second moment	$\sum_{i=1}^{g}\sum_{j=1}^{g} p^2(i,j,d,\theta)$	The uniformity of gray distribution and degree of texture.
T_2	Sums of average	$\sum_{k=2}^{2g} k \times P_X(k)$	The change of brightness.
T_3	Sums of variance	$\sum_{k=2}^{2g} (k-T_2)^2 P_X(k)$	Texture period size.
T_4	Maximum probability	$\underset{i,j}{MAX}[p(i,j,d,\theta)]$	The distribution of the main texture.
T_5	Sums of entropy	$-\sum_{k=2}^{2g} P_X(k) \times \log[P_X(k)]$	Texture complexity.
T_6	Variance	$\sum_{i=1}^{g}\sum_{j=1}^{g} (i-m)^2 p(i,j,d,\theta)$	Texture periodicity.
T_7	Variance of grayscale	$\sum_{i=1}^{g}\sum_{j=1}^{g} [(i-j)^2 \times p^2(i,j,d,\theta)]$	Texture distribution.
T_8	Correlation	$\sum_{i=1}^{g}\sum_{j=1}^{g} [(i \times j \times p(i,j,d,\theta) - u_1 \times u_2] / (d_1 \times d_2)$	The main direction of the texture.
T_9	Inverse matrix	$\sum_{i=1}^{g}\sum_{j=1}^{g} p(i,j,d,\theta)/[1+(i-j)^2]$	Local texture changes.
T_{10}	Cluster shadow	$\sum_{i=1}^{g}\sum_{j=1}^{g} [(i-u_1)+(j-u_2)]^3 \times p(i,j,d,\theta)$	Texture uniformity.
T_{11}	Significant clustering	$\sum_{i=1}^{g}\sum_{j=1}^{g} [(i-u_1)+(j-u_2)]^4 \times p(i,j,d,\theta)$	Texture uniformity.
T_{12}	Entropy	$\sum_{i=1}^{g}\sum_{j=1}^{g} p(i,j,d,\theta)^2 \times \log_{10} p(i,j,d,\theta)$	Texture randomness.

Note: g takes a power of 2, $u_1 = \sum_{i=1}^{g} i \sum_{j=1}^{g} p(i,j,d,\theta)$, $u_2 = \sum_{j=1}^{g} j \sum_{i=1}^{g} p(i,j,d,\theta)$, $d_1^2 = \sum_{i=1}^{g} (i-u_1)^2 \sum_{j=1}^{g} p(i,j,d,\theta)$, $d_2^2 = \sum_{j=1}^{g} (j-u_1)^2 \sum_{i=1}^{g} p(i,j,d,\theta)$, $T_2 = \sum_{k=2}^{2g} k \times P_X(k)$, m is the average of $p(i,j,d,\theta)$, $P_x(k) = \sum_{i=1}^{g}\sum_{j=1}^{g} p(i,j,d,\theta) \Big|_{|i+j|=k}$, k is $2, 3, \ldots, 2g$.

2.3. SOM Networks Model

The self-organizing map neural network (SOM); strengthening central neurons and inhibiting peripheral neurons is a unique property of biological neuronal systems. The patterns of competition between neurons is a mechanism formed by the brain's self-organization [37]. SOM retains the spatial topology of input neurons due to its unique neighborhood characteristics and obtains the best mapping by continuously adjusting the connection weights of winning neurons and neighboring neuron network nodes [38]. The SOM neural network closely combines the phenomenon of enhancing the center and suppressing the surrounding of the biological neuron system with the self-organization characteristics of the human brain. Therefore, SOM retains the topology of the network and can be based on limited sample space failure detection, effectively reducing the need for spatial samples in the intelligent diagnostic system. Figure 2 shows a typical self-organizing feature mapping network model consisting of an input layer and a contention layer. The input layer contains n neurons, and the competition layer is a two-dimensional planar array composed of $m \times n$ neurons [39]. This section utilizes SOM to identify micro-damage structures.

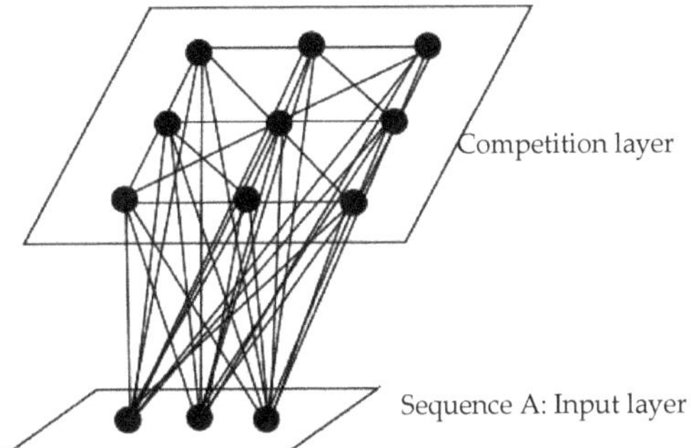

Figure 2. The Self-Organizing Mapping (SOM) network model.

2.4. Learning Steps

(i) Initialization: Set the [0,1] random value as the initial connection weight between the input neuron and the output neuron. A set, S_j, of outputting neighboring neurons is selected, wherein $S_j(0)$ represents a set of neighboring neurons of neuron j at time $t = 0$, and $S_j(t)$.

(ii) Set the input of the neural network: Make the sample feature parameters into the following matrix and input them to the SOM network:

$$X(n) = [x_1(n), x_2(n), x_3(n) \cdots x_{N-1}(n), x_N(n)]^T. \tag{3}$$

(iii) Calculate the Euclidean distance: Input layer neurons, i, into mapping layer neurons', j, available Euclidean distance, d_{ij}, indicated as:

$$d_{ij} = \|X - W_j\| = \sqrt{\sum_{i=1}^{n} [x_i(t) - w_{ij}(t)]^2}. \tag{4}$$

In the equation, w_{ij} is the weight of the input layer neuron, i, to the mapping layer neuron, j. W_j is the connection weight of the neuron, j, on the mapping layer.

(iv) Obtain the winning neuron: The position of the winning neuron can be obtained by calculating the minimum Euclidean distance between the input vector and the weight vector. When the input vector is X and the winning neuron is denoted by c, the formula is expressed as:

$$\|X - W_c\| = \min_i \|X - W_i\|, \; i = 1, 2, 3 \cdots M - 1, M, \tag{5}$$

where x is the input vector and the winning neuron is labeled c. W_c is the weight of the winning neuron, c. W_i is the connection weight of the neuron, i, on the mapping layer.

(v) Adjust weight: The connection weight of the input neuron and all neurons in the competition neighborhood are corrected by Equation (6):

$$\Delta w_{ij} = w_{ij}(t+1) - w_{ij}(t) = \eta(t)[x_i(t) - w_{ij}(t)]. \tag{6}$$

Among them, t is the continuous time, and the learning rate at time t is $\eta(t)$. $\eta(t) = \frac{1}{t}$ or $\eta(t) = 0.2(1 - \frac{t}{1000})$. The value range of $\eta(t)$ is [0,1].

(vi) Determine whether the output result meets the expected requirements: If the result meets the previously set requirements, then end; if not, return to step (ii) to continue.

3. Experiment and Result Analysis

A diagnosis model based on SOM and GLCM was established in order to obtain the results of the micro-damage recognition of underwater structures. The diagnosis steps are shown in Figure 3.

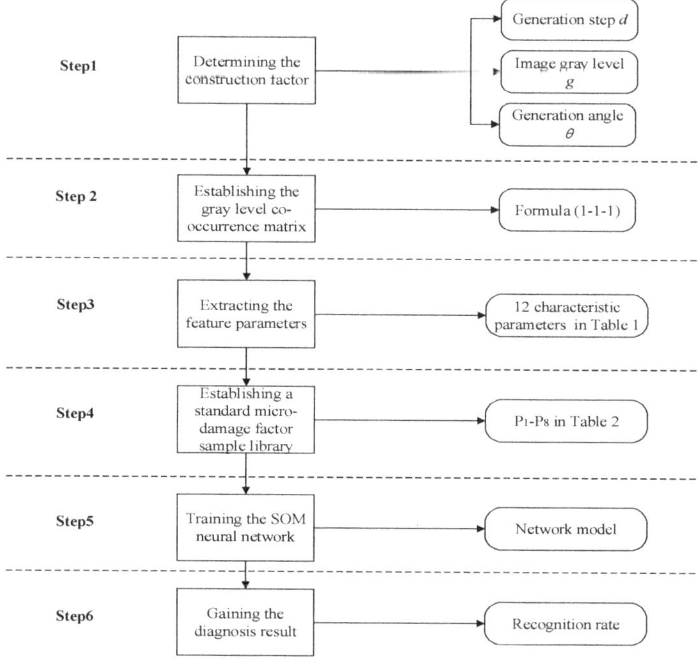

Figure 3. The diagram of the diagnostic process.

As shown in Figure 3, the diagnostic process consists of six steps. The diagnosis process is detailed on the left side, including the generation of grayscale co-occurrence matrices, the establishment of standard samples, the training of network models, and the acquisition of diagnostic results. The right side of Figure 3 is the information about the important parameters or results corresponding to the diagnostic steps.

3.1. Image Acquisition and Processing

An underwater micro-damage visual diagnosis system was established according to the characteristics of underwater structures, as shown in Figure 4. The system mainly contains a HCCK-102 series crack detection system (including industrial camera, industrial endoscope, control interface), maintenance pool, concrete test block, computer etc. Among them, the industrial endoscope has a magnification of 40x and a measurement accuracy of 0 to 0.08 mm. The micro-damage image of the underwater concrete test block was obtained based on the diagnosis system. The image definition was extremely reduced due to the turbid water quality, insufficient light, and the configuration of the protective cover for the micro-probe. Underwater micro-damage images were processed by ROI (region of interesting) interception, histogram equalization, and a median filter.

Figure 4. Acquisition system.

3.2. Triangle Algorithm

The accuracy of extracting feature parameters depends on the optimal generation criteria, including the image gray level, g, the generative step length, d, and the generative angle, θ. Therefore, it is crucial to optimize the factor to obtain accurate and effective feature parameters. Generally, the orthogonal test method (OTM) and full test method (FTM) are used to establish generative criterion. However, OTM leads to the absence of a partial criterion and FTM reduces detection efficiency. Therefore, a new triangulation algorithm used for the confirming of generative criterion is presented in this paper. Figure 5 is the diagram of algorithm.

Step 1: The initial range of the factor of the generative criterion is determined according to the properties of the micro-damage image.

Step 2: Confirm the generative angle, θ, by the theory of image rotation invariant. In Figure 5, the *sequence A* is he generative step length, d, g_1-g_{nn} is the *sequence B* image gray level, g, g_1-g_n.

Step 3: The sequence A is linked to the sequence B, and θ is then joined to sequence A and sequence B, respectively.

Step 4: Extract all d_n-θ-g_t combinations. Then, a triangular combination can form.

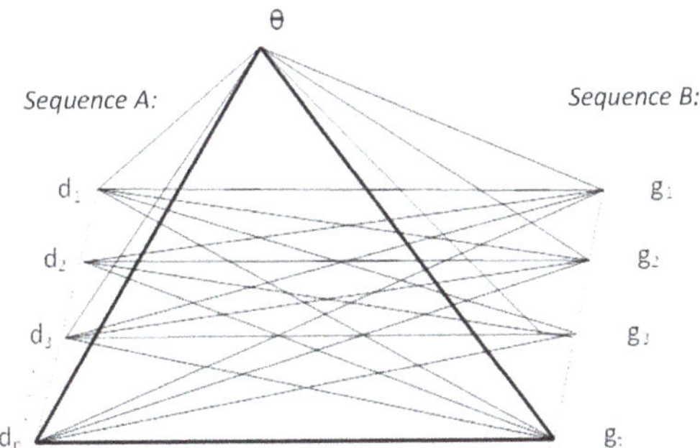

Figure 5. Algorithm.

The algorithm was validated based on the underwater micro-damage diagnosis system. Based on the comparative analysis, the unique advantages of the triangulation algorithm were proven.

In order to ensure the features of the image rotation invariant, the average values directions of $0°$, $45°$, $90°$, and $135°$ were selected as the generative angle, θ. Figure 6 is the variations' form of θ.

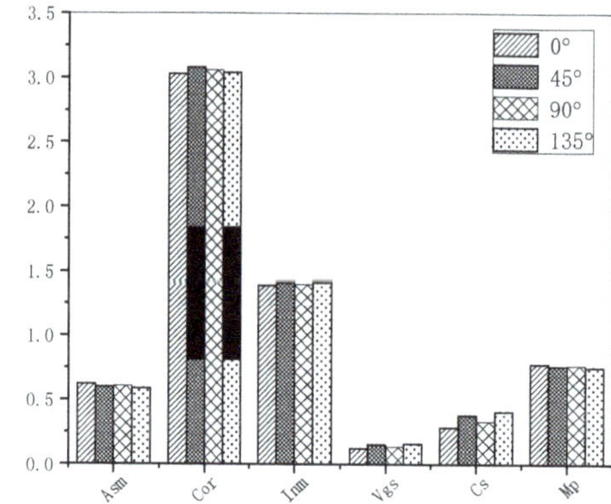

Figure 6. Variations form of θ.

Figure 6 shows the difference in the characteristic parameters of micro-damage images. The different directions are small. Therefore, the generation angle, θ, of the generation criterion was determined.

The initial range of the generated standard factor was obtained based on the characteristics of the micro-damage image:

(1)   Generative angle, θ: The average of directions of 0°, 45°, 90°, and 135°.
(2)   Image gray level: $g_t = 2^{m+2}$, among them, $t = m + 2$, t is taken as the integer of [1,6].
(3)   Generative step length, d: Take the integer of [1,6].

The combination form of the generation standard, dn-θ-gt, was obtained by the triangulation algorithm. It has a total of $n \times t = n \times (m + 2) = 36$ groups. The OTM requires determination of the values of the two construction factors in advance to obtain the combined form. If the two construction factors select g and d, the d-θ-g forms a total of $1 \times 1 \times 4 = 4$ groups. Additionally, the FTM does not specify the range of variation of any factor. If the values of d and g are the same as the triangulation algorithm, the d-θ-g forms a total of $6 \times 6 \times 4 = 144$ groups. Experimental verification and comparison results showed that this method can efficiently obtain all the generation criteria of micro damage. The proposed triangulation algorithm obtained the appropriate combination form based on the image properties and effectively improved the detection efficiency.

3.3. Optimization of Generative Criterion

To improve the characteristic performance of structure damage, the generative criterion optimization of GLCM was carried out based on the theory of difference maximization. Figure 7 show the variation tendency of d, in which the selected characteristic parameters showed a good correlation with the change trend of d. Figure 8 shows the variation form of g. The selected characteristic parameters are sensitive to the changing trend of g.

It can be derived that when $d > 4$, the characteristic parameter changes tend to be stable basically; when $d = 5$, the samples of each group can be better distinguished; when $d = 6$, there is a clear downward trend in aggregation in the correlation, the significant clustering, and the clustering shadow characteristic parameters.

Figure 8 is the variation of g of six characteristic parameters, including angular second-order moment, entropy maximum probability, etc. It can be analyzed that when $g = 5$, each characteristic parameter exhibits high discreteness and high stability.

All the analyses show that the optimal generative criterion, d_n-θ-g_t, of the underwater structure micro-damage is n is 5, t is 7, and θ is taken as the average of the four directions of 0°, 45°, 90°, and 135°.

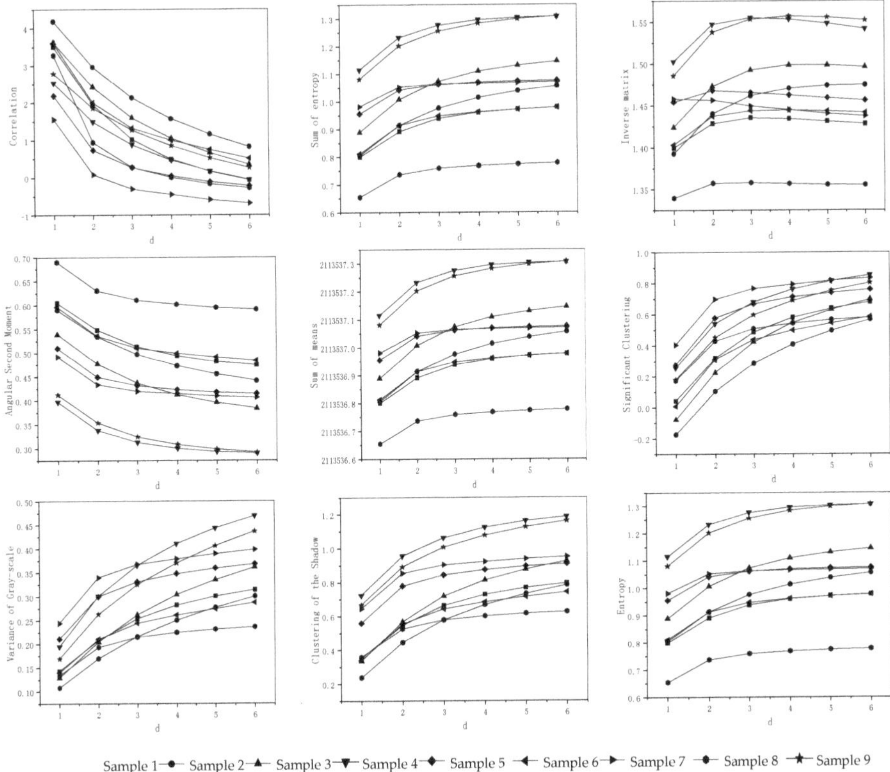

Figure 7. Tendency of d (when $g = 2^8$). Sample 1–9 represents nine sample images of certain types of defects.

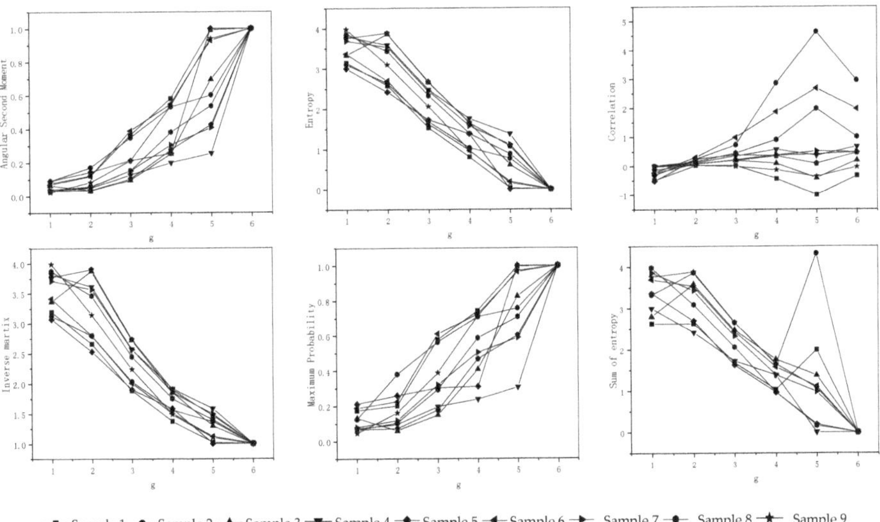

Figure 8. Form of g (when $d = 5$). Sample 1–9 represents nine sample images of certain types of defects.

3.4. Establishing a Standard Sample Label

For the purpose of obtaining the standard performance parameters of the SOM diagnostic model, the standard sample label was established using the DFS (digital feature screening) method under the optimal generative criteria (d_5-θ-g_7, among them, θ takes 0°, 45°, 90°, and 135°). The DFS method establishes screening criteria for screening parameters by observing the trend of the characteristic parameters of different diagnostic types. The screening criteria are determined according to the size of the intersection interval of different types of damages: The smaller the crossover interval, the better the screening results, and the parameters are suitable for establishing a diagnostic model; otherwise, the parameter is not suitable for diagnosing underwater structural damage. Figures 9–11 shows the screening results.

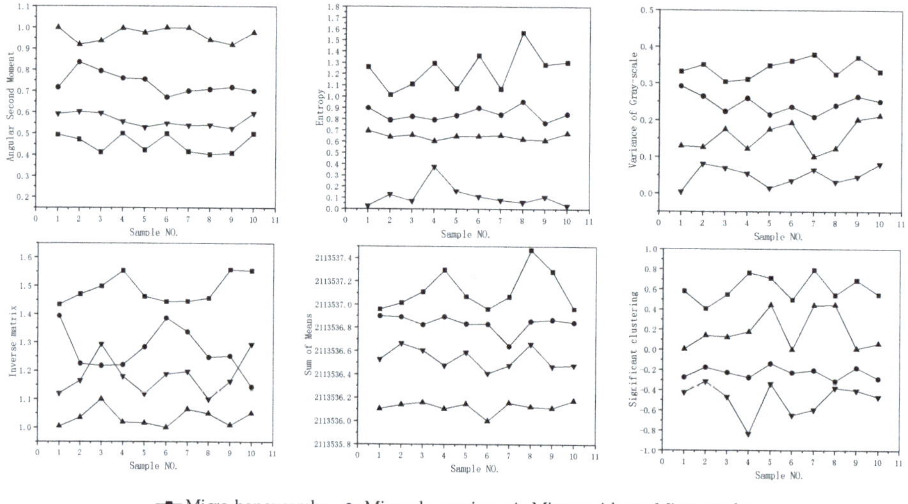

Figure 9. Parameters with preferable screening results.

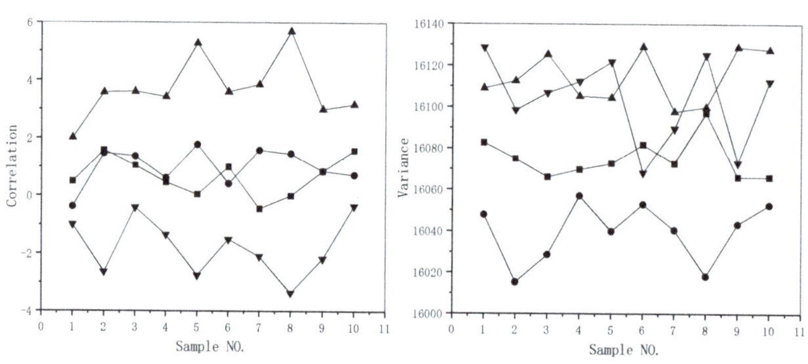

Figure 10. Parameters with ordinary screening results.

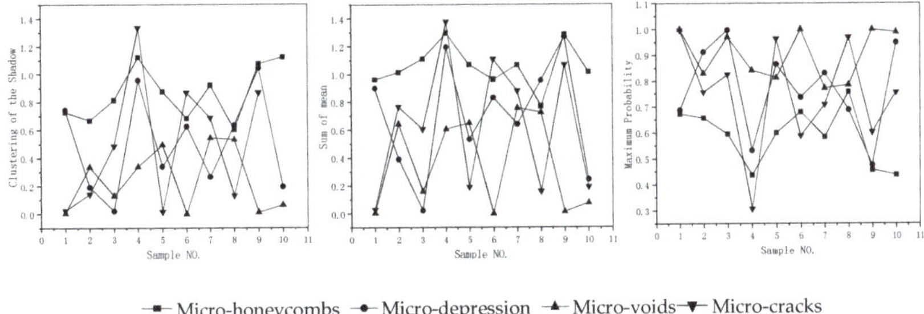

–■– Micro-honeycombs –●– Micro-depression –▲– Micro-voids –▼– Micro-cracks

Figure 11. Parameters with inferior screening results.

In Figure 9, the characteristic parameters occupy their respective intervals with preferable screening results. In Figure 10, the two types of damage have a small interaction interval, the screening results are ordinary. In Figure 11, each of the three characteristic parameters is interlaced with each other and each has no independent interval.

The angular second moment, P_1, entropy, P_2, variance of grayscale, P_3, correlation, P_4, inverse matrix, P_5, variance, P_6, significant clustering, P_7, and sums of mean, P_8, were screened. Therefore, the DFS method could effectively solve these problems of characteristic parameter selection. The standard values of each parameter are given in Table 2.

Table 2. Normal sample.

Damage Type	P_1	P_2	P_3	P_4	P_5	P_6	P_7	P_8
Micro-honeycombs	0.45219	1.23347	0.341173	0.659015	1.487459	16074.99305	0.6085123	2113537.15
Micro-depressions	0.735626	0.844082	0.245272	0.986124	1.270514	16039.67163	−0.23044	2113536.83
Micro-voids	0.966008	0.645785	0.155829	3.719458	1.034531	16113.92133	0.185468	2113536.14
Micro-cracks	0.561931	0.114353	0.047756	−1.768837	1.1816	16103.60313	−0.487734	2113536.55

Table 2 gives the standard value of the characteristic parameters corresponding to each kind of damage, which were obtained according to the average value of 200 images.

3.5. Training Network Model

The network model was constructed. The parameters of the network are shown in Table 3.

Table 3. Network parameters.

Parameters	Value							
Input Layer Node	x_1	x_2	x_3	x_4	x_5	x_6	x_7	x_8
	P_1	P_2	P_3	P_4	P_5	P_6	P_7	P_8
Weight	0.125							
Neighborhood shape	Hexagon, R = 3							
Neurons	64							
Training steps	10, 50, 100, 200, 500,1000							

The created network topology is a two-layer network, and the competition layer consists of eight vectors. $[x_1, x_2, x_3, x_4, x_5, x_6, x_7, x_8]$. The initial weight was set to 0.125. The neighborhood shape of each network node was set to a hexagon, and the neighborhood radius was set to 3. The competition layer was 8 × 8 = 64 neuron nodes. The training steps were set to 10, 50, 100, 200, 500, and 1000. Table 4 shows the classification effects under different training steps; the numbers in the table represent the classification numbers.

Table 4. Classification results for different steps.

Number of Training Steps	Micro-Honeycombs	Micro-voids	Micro-depressions	Micro-cracks	Clustering Result
10	55	37	37	55	50%
50	43	37	37	55	75%
100	43	1	37	37	75%
200	49	1	16	64	100%
500	49	1	16	64	100%
1000	49	1	16	64	100%

Table 4 shows that when the number of training steps equals 10, a micro-damage diagnosis model is initially established, and the micro-damage is roughly divided into two groups. With the increase of the number of training steps, when the number of training steps is 50 and 100, respectively, the diagnostic accuracy is further improved. The micro-injury is divided into three groups. When the number of training steps reaches 200, the micro-injury is effectively distinguished. However, when the number of training steps reaches 500 and 1000, the micro-injury classification results are the same as 200 steps, which has no practical significance. Therefore, 200 steps 200 were chosen as the optimal value for the micro-damage diagnosis model.

To verify the reliability of parameter optimization, the topological structure of the winning neurons in Figure 12 was analyzed. The gray hexagon with the number in the figure corresponds to the classification label of the winning neurons.

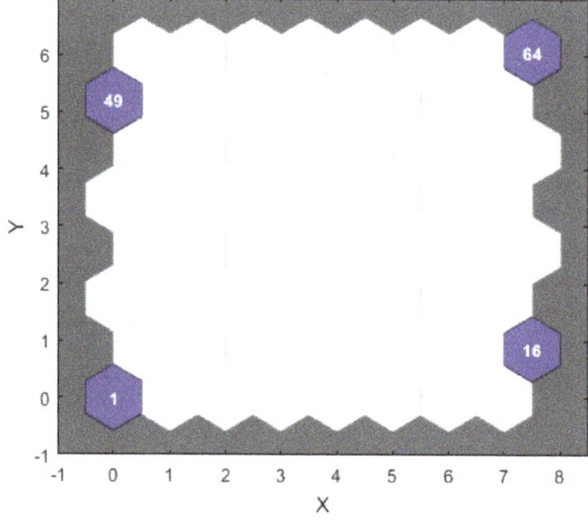

Figure 12. Winning neuron topology.

The unique neighborhood characteristics of SOM and the spatial topology of the input neurons are preserved. Figure 12 shows that the winning neuron numbers are 1, 16, 49, and 64, occupying four different topological spatial locations. The classification result of the topological result graph is completely consistent with the clustering result obtained under the optimal parameters of the network. By analyzing the topology, the network parameters were optimized to build a better SOM network model. It is proven that the setting of various parameters of the network can meet the working requirements of an underwater structure diagnosis system and effectively distinguish the types of micro-damage.

Although the topological position of the winning neurons was obtained, it cannot determine the type of damage for each neuron. Therefore, the neuron distance map was obtained as shown in Figure 13. Small square hexagons represent neurons and straight lines represent linear connections between neurons. The distance between neurons can be derived from the Euclidean distance formula.

The long hexagon of the connected neurons and the depth of the color represents the distance between the neurons.

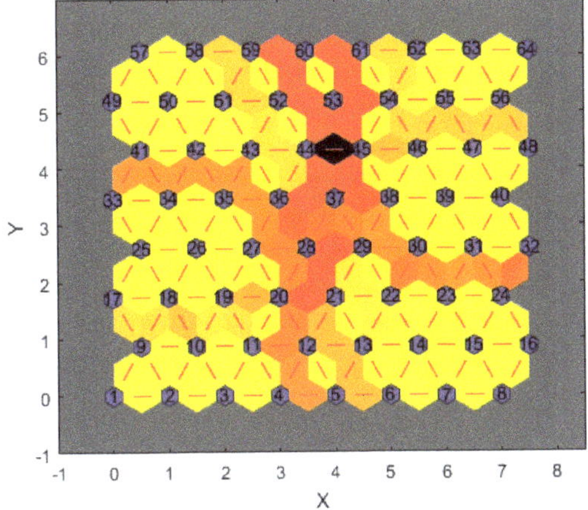

Figure 13. Distance between the neurons.

In Figure 13, the connected region of the dark hexagon divides the whole neuron into four sub-regions, corresponding to different damage types. The neurons of the light-colored hexagons connected in the sub-area have the same type of damage. Neurons 53 and 60 are surrounded by dark hexagons, corresponding to unknown damage types. Neuron 28 is far from the four types of damage and is classified as micro-voids defects with close proximity. Alternatively, it can be classified as other unknown defects different from 53 and 60 or 28. Therefore, the corresponding situation of all neuronal damage is obtained.

In order to determine the efficiency of the algorithm, test results were analyzed under different test conditions. Table 5 shows the classification of damage.

Table 5. Classification of damage.

Damage Type	Sample Classification Number	Classification Accuracy
Micro-honeycombs	36, 41, 42, 43, 44, 49, 50, 51, 52, 57, 58, 59	80%
Micro-voids	1, 2, 3, 4, 9, 10, 17, 18, 19, 20, 25, 26, 27, 33, 34, 35	93.33%
Micro-depressions	5, 6, 7, 8, 11, 12, 13, 14, 15, 16, 21, 22, 23, 24, 29	100%
Micro-cracks	30, 31, 32, 38, 39, 40, 45, 46, 47, 48, 54, 55, 56, 61, 62, 63, 64	86.66%
Unknown situation	53, 60	-
	28	-
	37	-

Table 5 describes the types of damage that the damage model can identify and the recognition rate for identifying the four types of damage. The data in Table 5 show that the proposed algorithm can not only detect the expected type of defect but also diagnose other unknown defects in the underwater concrete. It can be inferred that when the concrete structure is healthy, Figure 11 will consist of hexagons of a uniform color. Its classification accuracy, R, can be calculated by the formula: $R = \frac{N_{all} - N_{ab}}{N_{all}} \times 100\%$. While there are only four kinds of damage in the structure, each damage type can be accurately identified. The algorithm can also accurately extract unknown defects. Therefore, the algorithm can obtain good detection results in the case of correct detection, false positive, and false negative detection. It is proved that the algorithm can be applied to the actual detection of underwater structural damage.

3.6. Application and Validation

In order to prove the practicability and reliability of the network parameters and indicators, this method was applied as an example in a measurement project. Figure 14 shows a schematic diagram of the image acquisition system. The micro-damage images obtained are shown in Figure 15.

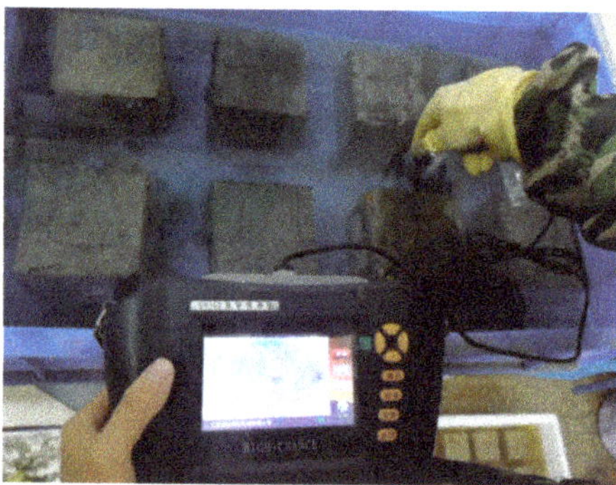

Figure 14. Image acquisition.

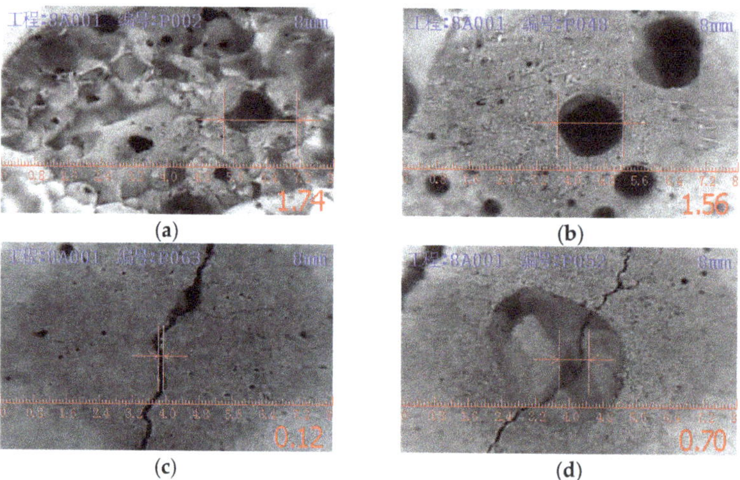

(a)

(b)

(c)

(d)

Figure 15. Micro-damage image. (**a**) Micro-honeycombs; (**b**) Micro-voids; (**c**) Micro-cracks; (**d**) Micro-depressions.

A test sample was constructed as shown in Table 6. The case of the predicted classification of the sample is shown in Figure 16.

Table 6. Damage prediction samples.

Damage Type	Sample Number
Micro-honeycombs	1–10
Micro-voids	11–20
Micro-depressions	21–30
Micro-cracks	31–40

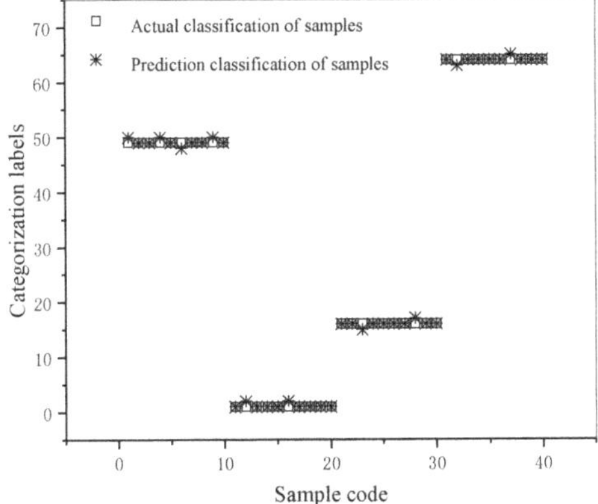

Figure 16. Sample prediction classifications.

This paper used the research methods of comparative analysis and graphs analysis. We can draw the conclusion that each classification predictive sample can correspond to the damage state. All prediction samples were correctly classified and the overall classification accuracy rate was 100%. It showed that the intelligent early warning method achieved satisfactory results and can be used as an intelligent and accurate diagnosis algorithm for solving underwater structural damage under complex constraints.

4. Conclusions

A vision-based approach for the diagnosis of damage on underwater structures was proposed by combining GLCM and SOM.

The underwater structural micro-damage images database required for training, validation, and application were conducted with NDT (non-destructive testing) equipment. To secure a wide range of adaptability, the damage images were acquired from the database randomly.

A triangle algorithm was proposed to generative criterion optimization, which has extracted comprehensive aggregation for the GLCM constructed. The generative criterion was confirmed based on the difference maximization principle: d_5-θ-g_7, which was the texture average of distance, delta = 4, four directions (0°, 45°, 90°, 135°), and image gray level, $g = 128$.

The SOM standard performance parameters were obtained by applying the DFS method, including the angular second moment, entropy, variance of grayscale, correlation inverse matrix, variance, significant clustering and sums of mean. The numbers of layers, numbers of nodes in hidden layers, and learning steps were set to optimal values by training the SOM network model and analyzing the SOM topology diagram. The measurement project example proved the robustness and adaptability of the method. The application result showed high performance under complicated conditions, including adverse environmental conditions, limited sample space, and complex defect types.

The results show that the proposed method exhibited good performance and can indeed diagnose structure damage in realistic underwater situations. In the future, intelligent warning techniques will replace human-conducted on-site inspections for the detection of underwater infrastructure. The method will also be combined with an unmanned aerial vehicle (UAV) to monitor the damage of civil structures.

Author Contributions: K.L. took part in the entire researching process. J.W. contributed to the experiment design. D.Q. revised the manuscript.

Funding: This research was funded by the Fundamental Research Fund for the Central University (Project No.2572018AB31), Heilongjiang Provincial Postdoctoral Scientific Research Development Fund (Project No.LBH-Q15011).

Conflicts of Interest: The authors declare no conflict of interest.

References

1. Farrar, C.R.; Worden, K. An introduction to structural health monitoring. *Philos. Trans. A Math. Phys. Eng. Sci.* **2010**, *365*, 303–315. [CrossRef] [PubMed]
2. Amezquita-Sanchez, J.P.; Adeli, H. Signal Processing Techniques for Vibration-Based Health Monitoring of Smart Structures. *Arch. Comput. Methods Eng.* **2016**, *23*, 1–15. [CrossRef]
3. Bhuiyan, M.Z.A.; Wang, G.; Wu, J.; Cao, J.; Liu, X.; Wang, T. Dependable Structural Health Monitoring Using Wireless Sensor Networks. *IEEE Trans. Dependable Secure Comput.* **2017**, *14*, 363–376. [CrossRef]
4. Park, J.Y.; Lee, J.R. Application of the ultrasonic propagation imaging system to an immersed metallic structure with a crack under a randomly oscillating water surface. *J. Mech. Sci. Technol.* **2017**, *31*, 4099–4108. [CrossRef]
5. Sidibe, Y.; Druaux, F.; Lefebvre, D.; Maze, F.L. Signal processing and Gaussian neural networks for the edge and damage detection in immersed metal plate-like structures. *Artif. Intell. Rev.* **2016**, *46*, 289–305. [CrossRef]
6. Abdeljaber, O.; Avci, O.; Kiranyaz, S.; Gabbouj, M.; Inman, J.M. Real-time vibration-based structural damage detection using one-dimensional convolutional neural networks. *J. Sound Vib.* **2017**, *388*, 154–170. [CrossRef]
7. Bhuiyan, M.Z.A.; Cao, J.; Wang, G. Deploying Wireless Sensor Networks with Fault Tolerance for Structural Health Monitoring. In Proceedings of the IEEE 8th International Conference on Distributed Computing in Sensor Systems, Hangzhou, China, 16–18 May 2012.
8. Zong, Z.; Zhong, R.; Zheng, P.; Qin, Z.; Liu, Q. Research Progress and Challenges of Bridge Structure Damage Prognosis and Safety Prognosis Based on Health Monitoring. *China J. Highw. Transp.* **2014**, *27*, 46–57.
9. Alavi, A.H.; Hasni, H.; Jiao, P.; Borchani, W.; Lajnef, N. Fatigue cracking detection in steel bridge girders through a self-powered sensing concept. *J. Constr. Steel Res.* **2017**, *128*, 19–38. [CrossRef]
10. Rutkowski, T.A.; Prokopiuk, F. Identification of the Contamination Source Location in the Drinking Distribution System Based on the Neural Network Classifier. *IFAC-PapersOnLine* **2018**, *51*, 15–22. [CrossRef]
11. Chatterjee, S.; Sarkar, S.; Hore, S.; Dey, N.; Ashour, A.S.; Balas, V.E. Particle Swarm Optimization Trained Neural Network for Structural Failure Prediction of Multi-storied RC Buildings. *Neural Comput. Appl.* **2016**, *28*, 2005–2016. [CrossRef]
12. Xu, K.; Deng, Q.; Cai, L.; Ho, S.; Song, G. Damage Detection of a Concrete Column Subject to Blast Loads Using Embedded Piezoceramic Transducers. *Sensors* **2018**, *18*, 1377. [CrossRef] [PubMed]
13. Feng, D.; Feng, M.Q. Experimental validation of cost-effective vision-based structural health monitoring. *Mech. Syst. Signal Process.* **2017**, *88*, 199–211. [CrossRef]
14. Gao, W.; Zhang, G.; Li, H.; Huo, L.; Song, G. A novel time reversal sub-group imaging method with noise suppression for damage detection of plate-like structures. *Struct. Control Health Monit.* **2017**, *25*, e2111. [CrossRef]
15. Cha, Y.J.; Choi, W.; Suh, G.; Mahmoudkhani, S.; Büyüköztürk, O. Autonomous Structural Visual Inspection Using Region-Based Deep Learning for Detecting Multiple Damage Types. *Comput. Aided Civ. Infrastruct. Eng.* **2017**, *33*, 731–747. [CrossRef]
16. Zhu, Z.; German, S.; Brilakis, I. Visual retrieval of concrete crack properties for automated post-earthquake structural safety evaluation. *Autom. Constr.* **2011**, *20*, 874–883. [CrossRef]
17. Molero, M.; Aparicio, S.; Al-Assadi, G.; Casati, M.J.; Hernandez, M.G.; Anaya, J.J. Evaluation of freeze–thaw damage in concrete by ultrasonic imaging. *NDT E Int.* **2012**, *52*, 86–94. [CrossRef]
18. German, S.; Brilakis, I.; Desroches, R. Rapid entropy-based detection and properties measurement of concrete spalling with machine vision for post-earthquake safety assessments. *Adv. Eng. Inform.* **2012**, *26*, 846–858. [CrossRef]
19. Hasni, H.; Alavi, A.H.; Jiao, P.; Lajnef, N. Detection of fatigue cracking in steel bridge girders: A support vector machine approach. *Arch. Civ. Mech. Eng.* **2017**, *17*, 609–622. [CrossRef]
20. Vetrivel, A.; Gerke, M.; Kerle, N.; Nex, F.; Vosselman, G. Disaster damage detection through synergistic use of deep learning and 3D point cloud features derived from very high resolution oblique aerial images, and multiple-kernel-learning. *ISPRS J. Photogramm. Remote Sens.* **2017**, *140*, 45–59. [CrossRef]
21. Yan, Y.; Cheng, L.; Wu, Z.; Yam, L.H. Development in Vibration-Based Structural Damage Detection Technique. *Mech. Syst. Signal Process.* **2007**, *21*, 2198–2211. [CrossRef]

22. Rafiei, M.H.; Adeli, H. A novel unsupervised deep learning model for global and local health condition assessment of structures. *Eng. Struct.* **2018**, *156*, 598–607. [CrossRef]

23. Cha, Y.J.; Choi, W.; Büyüköztürk, O. Deep Learning-Based Crack Damage Detection Using Convolutional Neural Networks. *Comput. Aided Civ. Infrastruct. Eng.* **2017**, *32*, 361–378. [CrossRef]

24. Liu, L.; Fieguth, P.; Guo, Y.; Wang, X.; Pietikäinen, M. Local binary features for texture classification: Taxonomy and experimental study. *Pattern Recognit.* **2017**, *62*, 135–160. [CrossRef]

25. Malegori, C.; Franzetti, L.; Guidetti, R.; Casiraghi, E.; Rossi, R. GLCM, an image analysis technique for early detection of biofilm. *J. Food Eng.* **2016**, *185*, 48–55. [CrossRef]

26. Maddalena, L.; Petrosino, A. A Self-Organizing Approach to Background Subtraction for Visual Surveillance Applications. *IEEE Trans. Image Process.* **2008**, *17*, 1168–1177. [CrossRef]

27. Li, X.; Zhu, D. An Adaptive SOM Neural Network Method to Distributed Formation Control of a Group of AUVs. *IEEE Trans. Ind. Electron.* **2018**, *65*, 8260–8270. [CrossRef]

28. Merainani, B.; Rahmoune, C.; Benazzouz, D.; Ould-Bouamama, B. A novel gearbox fault feature extraction and classification using Hilbert empirical wavelet transform, singular value decomposition, and SOM neural network. *J. Vib. Control* **2017**, *24*, 2512–2531. [CrossRef]

29. Li, Y.; Yao, X.; Li, W.; Li, C. Model Optimization of Wood Property and Quality Tracing Based on Wavelet Transform and NIR Spectroscopy. *Spectrosc. Spectr. Anal.* **2018**, *38*, 1384–1392.

30. Kamruzzaman, M.; Elmasry, G.; Sun, D.W.; Allen, P. Non-destructive assessment of instrumental and sensory tenderness of lamb meat using NIR hyperspectral imaging. *Food Chem.* **2013**, *141*, 389–396. [CrossRef]

31. Raju, P.; Rao, V.M.; Rao, B.P. Optimal GLCM combined FCM segmentation algorithm for detection of kidney cysts and tumor. *Multimed. Tools Appl.* **2019**, *78*, 18419–18441. [CrossRef]

32. Ancy, C.A.; Nair, L.S. Tumour Classification in Graph-Cut Segmented Mammograms Using GLCM Features-Fed SVM. *Intell. Eng. Inform.* **2018**, *695*, 197–208.

33. Chen, J.; Chen, Z.; Chi, Z.; Fu, H. Facial Expression Recognition in Video with Multiple Feature Fusion. *IEEE Trans. Affect. Comput.* **2018**, *9*, 38–50. [CrossRef]

34. Oliveira, R.B.; Papa, J.P.; Pereira, A.S.; Tavares, J.M.R.S. Computational methods for pigmented skin lesion classification in images: Review and future trends. *Neural Comput. Appl.* **2016**, *29*, 613–636. [CrossRef]

35. Torres-Alegre, S.; Fombellida, J.; Piñuela-Izquierdo, J.A.; Andina, D. AMSOM: Artificial metaplasticity in SOM neural networks—Application to MIT-BIH arrhythmias database. *Neural Comput. Appl.* **2018**, 1–8. [CrossRef]

36. Rodriguez-Galiano, V.F.; Chica-Olmo, M.; Abarca-Hernandez, F.; Atkinson, P.M.; Jeganathan, C. Random Forest classification of Mediterranean land cover using multi-seasonal imagery and multi-seasonal texture. *Remote Sens. Environ.* **2012**, *121*, 93–107. [CrossRef]

37. Rokach, L. A survey of Clustering Algorithms. *Data Mining Knowl. Discov. Handb.* **2009**, *16*, 269–298.

38. Vesanto, J.; Alhoniemi, E. Clustering of the self-organizing map. *IEEE Trans. Neural Netw.* **2000**, *11*, 586–600. [CrossRef]

39. Kohonen, T. Essentials of the self-organizing map. *Neural Netw.* **2013**, *37*, 52–65. [CrossRef]

Review

A Review of Lithium-Ion Battery Fault Diagnostic Algorithms: Current Progress and Future Challenges

Manh-Kien Tran and Michael Fowler *

Department of Chemical Engineering, University of Waterloo, Waterloo, ON N2L 3G1, Canada; kmtran@uwaterloo.ca
* Correspondence: mfowler@uwaterloo.ca; Tel.: +1-519-888-4567 (ext. 33415)

Received: 26 February 2020; Accepted: 6 March 2020; Published: 8 March 2020

Abstract: The usage of Lithium-ion (Li-ion) batteries has increased significantly in recent years due to their long lifespan, high energy density, high power density, and environmental benefits. However, various internal and external faults can occur during the battery operation, leading to performance issues and potentially serious consequences, such as thermal runaway, fires, or explosion. Fault diagnosis, hence, is an important function in the battery management system (BMS) and is responsible for detecting faults early and providing control actions to minimize fault effects, to ensure the safe and reliable operation of the battery system. This paper provides a comprehensive review of various fault diagnostic algorithms, including model-based and non-model-based methods. The advantages and disadvantages of the reviewed algorithms, as well as some future challenges for Li-ion battery fault diagnosis, are also discussed in this paper.

Keywords: lithium-ion battery; battery faults; battery safety; battery management system; fault diagnostic algorithms

---

1. Introduction

Lithium-ion (Li-ion) batteries play a significant role in daily applications due to their important advantages over other energy storage technologies, such as high energy and power density, long lifespan, and low self-discharge performance factors under improper temperatures [1]. Li-ion batteries have gained a significant amount of attention in recent years, showing promise as an energy storage source in electric vehicles (EVs) due to the aforementioned advantages [2]. They are also widely used in many electronics and stationary applications. Although Li-ion batteries have had reported accidents causing public concern, the advent of safety features over time has decreased the associated risk factors and improved battery operation [3,4]. Li-ion batteries have gone through many modifications to improve their safety, and system management is required to guarantee the performance and reliability of these batteries [1]. A key component in monitoring the battery health and safety is the battery management system (BMS). Some of its main functions are data acquisition, state of charge (SOC) and state of health (SOH) estimation, cell balancing, charge management, and thermal management. An important function of the BMS is the diagnosis of faults which can come from extreme operating conditions, manufacturing flaws, or battery aging [5].

Many factors affect the Li-ion battery operation, such as collision and shock, vibration, deformation, metallic lithium plating, formation of a solid electrolyte interphase (SEI) layer, formation of lithium dendrite, etc. [1,6]. Some common external battery faults are sensor faults, including temperature, voltage and current sensor faults, as well as cell connection and cooling system faults. There are also internal battery faults that are caused by the above factors and external battery faults. Some common internal battery faults are overcharge, overdischarge, internal and external short circuit, overheating, accelerated degradation, and thermal runaway. These battery faults lead to potentially hazardous consequences, such as an increase in temperature and pressure, which could increase the risk of combustion and explosion [7]. The consequences of Li-ion battery faults can be minimized or eliminated by the BMS, as it prevents the battery from functioning outside its safe operational range, and also detects faults using various fault diagnostic methods [8]. It is an indispensable component in the Li-ion battery system since it can handle failures safely by going into failure mode, providing a safer environment for the users of Li-ion battery applications.

Since fault diagnosis is a crucial part of Li-ion battery advancements, various methods for fault diagnosis have been studied and developed. Many fault diagnostic algorithms have been proposed, which can be categorized into model-based and non-model-based methods. Model-based methods include parameter estimation, state estimation, parity space, and structural analysis. Non-model-based methods include signal processing and knowledge-based methods [1,9]. Fault diagnosis research in other fields has shown that the most effective approach is often a combination of more than one method [9].

Lu et al. [8] briefly presented fault diagnosis as one of the key issues for Li-ion battery management in electric vehicles. In [10], a review of sensor fault diagnosis for Li-ion battery systems was provided, but other types of faults were not discussed. Wu et al. [1] conducted a review on fault mechanism and diagnosis for Li-ion battery, but there have been many new developments in the field since then. Researchers have made significant progress in understanding the mechanism and operation of Li-ion batteries, and with that, many innovations in battery fault diagnosis have emerged. Therefore, there is a need to identify the current progress and future direction of Li-ion battery fault diagnosis research. The contribution of this paper is the comprehensive review of recent developments in Li-ion battery fault diagnosis, as well as the discussion of some future challenges that need to be addressed.

The rest of this paper is organized as follows. Section 2 introduces different types of faults in Li-ion battery, and their causes, mechanisms, and consequences. The role of BMS in fault diagnosis is presented in Section 3. Section 4 provides a review of various fault diagnostic algorithms for the faults discussed in Section 2 and discusses the current status and future challenges of this subject. Finally, concluding remarks are given in Section 5.

2. Types of Fault in the Li-Ion Battery System

This paper categorizes Li-ion battery faults into internal and external faults. Lyu et al. [11] presented a detailed review of failure modes and mechanisms for rechargeable Li-based batteries. This section only briefly presents some causes and mechanisms of the faults. It is crucial to understand the mechanisms of these faults, as it helps establish appropriate fault diagnostic methods to ensure the safety of Li-ion battery applications. Figure 1 shows a summary of various faults in a Li-ion battery system.

2.1. Internal Battery Faults

Internal battery faults are difficult to detect since the operation within a Li-ion cell is still not fully understood [11]. Some examples of internal battery faults are overcharge, overdischarge, internal and external short circuit, overheating, accelerated degradation, and thermal runaway. All these faults affect the battery operation, but accelerated degradation and thermal runaway are the most dangerous since they can significantly affect the Li-ion battery application or directly harm the users [12]. Internal faults are often identified from abnormal responses from the battery operation, which include voltage drop, SOC drop, temperature rise, increase in internal resistance, and physical transformation, such as swelling. These responses are discussed further throughout this section.

2.1.1. Overcharge

Overcharge is a fault that can lead to more severe faults, such as accelerated degradation and thermal runaway. It may occur in Li-ion cells due to the capacity variation of cells in the pack, incorrect voltage and current measurement, or inaccurate SOC estimation from the BMS [13]. A normal battery pack can also get overcharged when the charger breaks down. Overcharging of Li-ion batteries leads to electrochemical reactions between battery components and the loss of active materials [14]. Additionally, in sealed batteries, the buildup of gases could cause the battery to burst [15]. Furthermore, the surface temperature of the battery increases significantly before it starts overcharging. This results in a thick SEI layer and also causes an internal short circuit inside the battery. The overcharged cathode suffers from electrolyte decomposition, metal dissolution, and phase transition, which could ultimately lead to thermal runaway and fires [16].

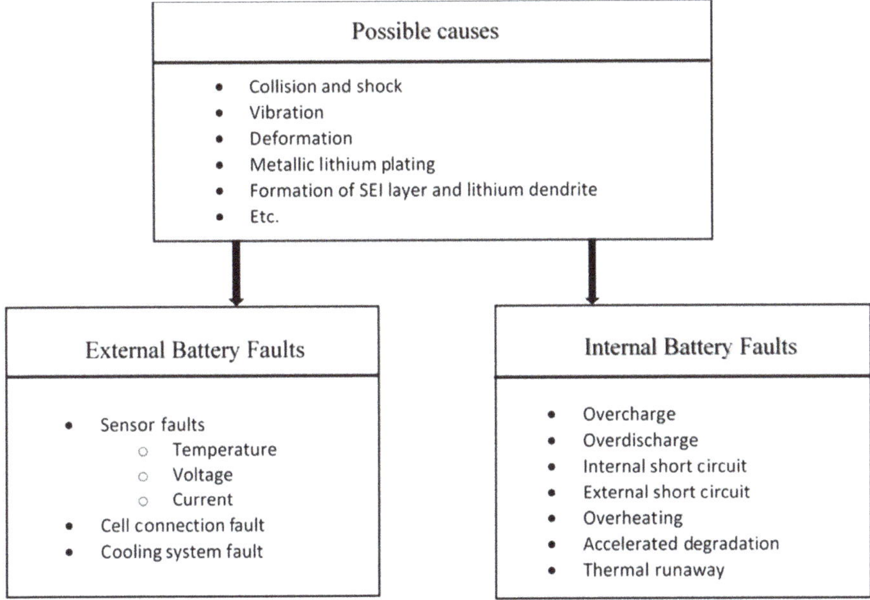

Figure 1. Internal and external Lithium-ion (Li-ion) battery faults and their causes.

2.1.2. Overdischarge

Overdischarge, similar to overcharge, can be caused by incorrect voltage and current measurements as well as inaccurate SOC estimation [13]. In [17], scanning electron microscopy and X-ray diffraction showed that overdischarge might be caused by copper deposition on the electrodes of the battery. Electrochemical impedance spectroscopy (EIS) studies also indicate that during battery overdischarge, the impedance of the anode is much smaller than that of the cathode, which means that the SEI change on the anode is much larger than the cathode, leading to capacity loss and current collection corrosion [14]. Overdischarge can impact the lifespan and thermal stability of a Li-ion cell, and result in considerable swelling of the cell [16]. The anode potential also increases abnormally during overdischarge, which can cause the Cu in the cell to oxidize to Cu^{2+} ions. The dissolved Cu^{2+} ions may result in migration through the separator and, ultimately, an internal short circuit [18].

2.1.3. Internal Short Circuit

A short circuit in Li-ion batteries can occur both externally and internally. An internal short circuit occurs when the insulating separator layer between the electrodes fails. This failure of the separator can be attributed to melting due to high temperature, cell deformation, the formation of dendrite, or compressive shock [7,19,20]. All of these can cause penetration through the separator layer, or cause the lithium ions and electrons to be released at the anode and travel across the electrolyte toward the cathode, which triggers contact between the anode and the cathode, leading to internal short-circuiting. When this phenomenon happens, the electrolyte tends to decompose by an exothermic reaction, causing thermal runaway [21]. Thermal runaway is mainly caused by heat build-up from the short circuit. In general, high capacity cells are at a higher risk of thermal runaway from internal short circuits than normal capacity cells [11].

2.1.4. External Short Circuit

An external short circuit generally occurs when the tabs are connected by a low resistance path [11]. Another cause is electrolyte leakage from cell swelling due to gas generation from side reactions during overcharge [22]. It can also occur due to water immersion and collision deformation. An external short circuit specifically transpires when an external heat-conducting element makes contact with the positive and negative terminals simultaneously, causing an electrical connection between the electrodes to occur [23]. A study [22] concluded that due to external short circuit, the Li-ion diffusion in the negative electrode leads to limited current, and the heat generated by the electrolyte decomposition in the positive electrode is responsible for thermal runaway occurrence. An external short circuit also leads to excessive discharge of the energy that is being stored in a cell [11].

2.1.5. Overheating

A Li-ion battery can overheat if an alternator's voltage regulator fails, sending a high amount of voltage back to the battery and causing overheating [24]. Overheating can also be caused by external and internal short circuits [25]. Overheating of the battery accelerates the degradation of the cathode and leads to SEI growth at the anode. As a result of overheating, there is a significant capacity loss. Overheating of Li-ion battery can lead to the materials inside the battery to break down and produce bubbles of gas, and, in most cases, the pressure build-up causes the battery to swell and possibly explode [26]. Another consequence of overheating is thermal runaway, which occurs because, at a critical temperature, a runaway reaction takes place as the heat cannot escape as rapidly as it is formed [15].

2.1.6. Accelerated Degradation

Cell degradation is a common characteristic in most batteries and occurs due to a variety of reasons, such as aging and self-discharging mechanisms. However, accelerated degradation is abnormal and can cause severe problems in Li-ion battery applications. The degradation process is accelerated during storage at elevated temperatures [27]. External degradation is also accelerated due to factors, such as impedance increase, higher frequency of cycle, change in SOC, and voltage rates [28,29]. Some mechanisms of accelerated degradation are corrosion of current collectors, changes in the electrode material, and reactions between electrodes and electrolyte. Accelerated degradation can cause the battery to have a shorter lifespan, which can be a major issue in applications such as EVs. It can also cause surface layer formation and contact deterioration, which results in electrode disintegration, material disintegration, and loss of lithium. These phenomena can lead to transport barriers, which result in penetration of the separator and cause an internal short circuit and, ultimately, a thermal runaway [30].

2.1.7. Thermal Runaway

All the above internal battery faults can cause thermal runaway. It can also occur during the charging of the battery under extreme charging currents and high temperatures. When the temperature reaches the melting point of the metallic lithium, it can cause a violent reaction [31]. Restricted air circulation is another cause of thermal runaway [32]. A study by Galushkin et al. [33] concluded that the probability of thermal runaway increases with the number of charge/discharge cycles. The study also found that thermal runaway is related to a variety of exothermic reactions in batteries. The first exothermic reaction to occur is SEI decomposition, and it considerably increases the heat release at the beginning of thermal runaway. During thermal runaway, the cathode releases oxygen by a phase transition, and the oxygen is consumed by the lithiated anode. A consequence of thermal runaway is a substantial increase in pressure and temperature of the Li-ion cell, which can lead to the destruction of the container and the release of a large amount of flammable and toxic gas [33]. Often in the case of a thermal runaway occurrence, the battery heats up and explodes. Hence, it is the most severe fault that can materialize in a Li-ion battery system.

2.2. External Battery Faults

External battery faults can have a significant effect on the other functions of the BMS and cause internal battery faults to occur. There are several types of external faults, which are temperature, voltage and current sensor faults, cell connection fault, and cooling system fault. The cooling system fault can be considered the most severe fault because it leads to a direct thermal failure, specifically thermal runaway, as the system fails to provide adequate cooling [34].

2.2.1. Sensor Fault

It is crucial to have a reliable sensor fault diagnostic scheme to ensure battery safety and performance. It also helps to prevent internal faults, such as overcharge, overdischarge, overheat, external and internal short circuit, and, most importantly, thermal runaway. Sensor faults include failure of temperature, voltage, and current sensors. Sensor faults are caused by vibration, collision electrolyte leakage, and other physical factors [35]. They can also be attributed to the loose battery terminals or corrosion around the battery sensor. A sensor fault can accelerate the degradation process of a battery, hinder the BMS functions due to incorrect state estimation, and cause other internal battery faults [10].

The temperature sensor is a key component in the Li-ion battery system, as it helps provide critical temperature data for the BMS to manage the battery operation effectively. A temperature sensor fault can cause it to send incorrect measurements to the BMS, which can cause further problems due to ineffective thermal management [36]. Inaccuracy in the BMS thermal management function can result in a significant decrease in battery life. Temperature sensor fault can also lead to short-circuiting, overheating, aging under high temperatures, capacity fade due to high temperature, and ultimately thermal runaway [33].

The voltage sensor is used to monitor the voltage of cells in the battery pack. A voltage sensor fault can cause the cell or the entire pack to exceed the upper and lower voltage limits that are specified by the manufacturers, which can result in overcharge and overdischarge [13]. A fault in the voltage sensor also leads to inaccurate SOC and SOH estimation, which can result in an internal fault as the battery suffers from overcharge and overdischarge [37].

The current sensor monitors the current that enters and exits the battery and sends the data to the BMS. It is important to detect a faulty current sensor as it can lead to further problems. The current can bypass the sensor, and the readings will be inaccurate [37]. A current sensor fault also leads to inaccurate estimation of SOC and other parameters, which impacts the control actions in the BMS, and causes the cell to overcharge, overdischarge, or overheat.

2.2.2. Cooling System Fault

The cooling system helps manage the thermal aspect of the battery. It transfers the heat away from the battery pack and ensures that the battery remains in the optimal temperature range. Cooling system faults occur when the cooling motor or fan fails to operate due to outdated fan wiring, faulty temperature sensor, or a broken fuse [2,38]. The temperature sensor and cooling system fault cannot be separated from each other as they both depend on a temperature range [34]. A cooling system fault is one of the more severe faults as it leads to a direct failure of the battery due to overheating, and ultimately thermal runaway. Therefore, it is important to diagnose it as early as possible.

2.2.3. Cell Connection Fault

Battery or cell connection fault is caused by the poor electrical connection between the cell terminals, as the terminals may become loose from vibration or corroded by impurities over time [39].

When this fault occurs, the cell resistance can increase drastically, leading to cell imbalance due to uneven current, or overheating of the faulty cell [40]. This type of fault is simple to detect with voltage and temperature sensors, but if left unresolved, it could lead to more severe consequences, such as external short circuit or thermal runaway.

3. The Role of BMS in Fault Diagnosis

One of the main functions of the BMS is to minimize the risks associated with the operation of a lithium-ion battery pack to protect both the battery and the users. Hazardous conditions are mostly caused by faults, and the safety functions of the BMS should minimize the likelihood of occurrence and the severity of these faults. Sensors, contactors, and insulation are common features added to the battery system to ensure its safety [13]. There are also operational limits for voltage, current, and temperature, which are monitored with sensors connected to the cells [41]. However, as the hardware and software implementation of the BMS becomes increasingly complex, battery faults can be more complicated, and these safety measures are often not adequate [42,43]. Fault diagnostic algorithms are, hence, a requirement for BMS. These algorithms serve the purpose of detecting faults early and providing appropriate and immediate control actions for the battery and the users [8]. Figure 2 illustrates the mechanism of fault diagnosis in the BMS.

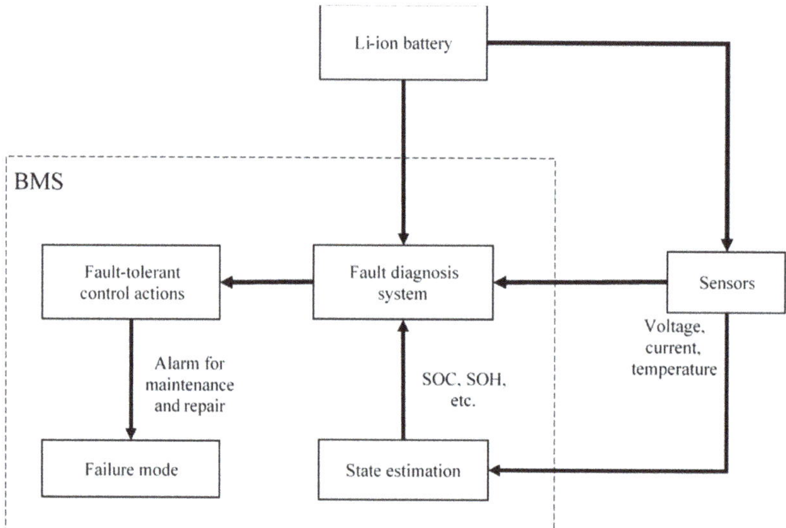

Figure 2. A schematic of fault diagnosis in the battery management system (BMS).

In the battery system, the BMS plays a significant role in fault diagnosis because it houses all diagnostic subsystems and algorithms. It monitors the battery system through sensors and state estimation, with the use of modeling or data analysis to detect any abnormalities during the battery system operation [13]. Since there are many internal and external faults, it is difficult to carry out this task efficiently. Various fault diagnostic methods need to work in tandem to correctly detect and isolate a specific fault, to administer the correct control action. However, the fault diagnostic algorithms in the BMS have limited computing space and time. Because of the large number of cells in the battery system in some applications, these fault diagnostic algorithms need to have low computational effort, while maintaining accuracy and reliability [44]. In recent years, there has been extensive effort in the research and development of efficient fault diagnostic approaches for Li-ion battery, which will be discussed in the following section.

4. Fault Diagnostic Algorithms for the Li-Ion Battery System

Fault diagnosis, as discussed, is an important function in the BMS. Fault diagnosis includes fault detection, isolation, and estimation. There are many fault diagnostic methods in various industries. For Li-ion battery applications, faults can be internal and interconnected, thus, many common methods in other fields are not suitable. There are two categories for fault diagnostic methods in Li-ion battery: model-based and non-model-based [9]. Classification of the fault diagnostic algorithms to be discussed in this paper can be found in Figure 3. This section presents recent developments for internal and external Li-ion battery fault diagnosis.

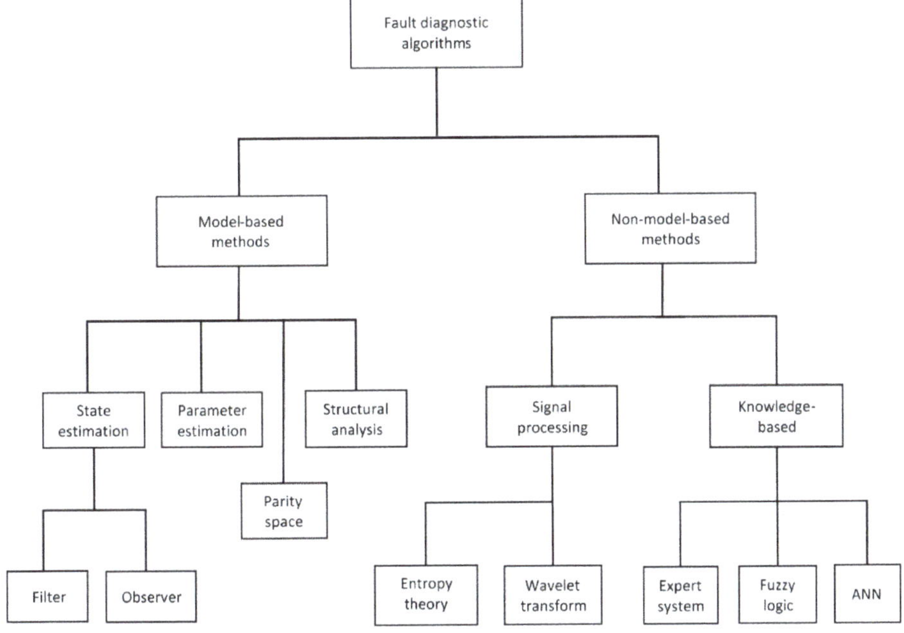

Figure 3. The classification of Li-ion battery fault diagnostic algorithms.

4.1. Internal Battery Fault Diagnosis

4.1.1. Model-Based Methods

The main principle of model-based fault diagnosis is the use of battery models to generate residuals which are monitored and analyzed to detect faults. There are several types of battery models, including electrochemical, electrical, thermal, and combinations of interdisciplinary models (electro-thermal, etc.) [45]. Each model can be used to assist fault diagnosis, depending on the requirements of the Li-ion battery application. Model-based methods are often used in fault diagnosis for their simplicity and cost-efficiency. Model-based methods include state estimation, parameter estimation, parity equation, and structural analysis. A simplified schematic for state estimation fault diagnostic algorithms is shown in Figure 4.

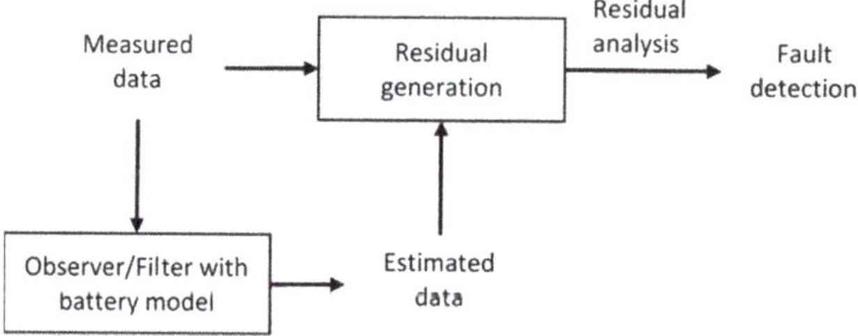

Figure 4. A simplified schematic of state estimation fault diagnosis.

In some earlier works, Alavi et al. [46] proposed a two-step state-estimation based algorithm for Li plating detection, which can lead to degradation and failure of the battery. The algorithm uses an electrochemical model, with the first step being estimating the insertion and extraction rates of Li ions using particle filtering, and the second step being the comparison of the estimated data with the boundary condition to generate alarms at faulty state. This method is effective in detecting Li plating, but the electrochemical model is complicated and not fully understood; hence, it is not reliable and impractical in the BMS. Authors in [47] used impedance spectroscopy to identify the parameters in the equivalent circuit model (ECM) to construct fault models for overcharge and overdischarge conditions. Kalman filters and the multiple-model adaptive estimation (MMAE) technique were proposed to generate residuals from the voltage, and the residuals were evaluated in a probability-based approach to detect overcharge and overdischarge faults. The same authors then proposed the use of extended Kalman filters with MMAE to detect overcharge and overdischarge faults [48]. This method eliminates noises effectively, but the use of multiple models makes it too complex and computationally inefficient. Chen et al. [49] used a bank of Luenberger observers with the ECM to generate residuals from voltage to locate and isolate faults in a series of three Li-ion cells and a bank of learning observers to estimate the isolated fault. This method was able to detect faults caused by changed internal resistance in a cell, but it is not practical for applications with a large number of cells, such as EVs.

In more recent works, the internal short circuit fault has been the main focus. Ouyang et al. [50] proposed an internal short circuit detection method using recursive least squares (RLS) to estimate the parameters of the mean-difference model, which is derived from the ECM. The parameters of all the cells in the pack are statistically analyzed and compared to a set threshold to detect internal short circuit fault. Authors in [51] used an electrochemical–thermal coupled model to confirm that an internal short circuit causes abnormal responses in voltage, temperature, and SOC. The ECM and the energy balance equation (EBE) model were used in the fault diagnostic algorithm instead of the electrochemical–thermal coupled model to lower the computational cost. RLS was utilized to estimate the parameters of these two models, and fault detection was achieved by observing changes in these parameters. The two diagnostic methods above are further developed in [52], where extended Kalman filters and RLS were implemented for state and parameter estimation, specifically SOC, voltage, temperature, and internal resistance. The state estimation and parameter estimation are performed on all the cells, with output data belonging to the mean and worst cells. Different scenarios of deviation from the estimated data indicate different levels of fault, including extra capacity depletion, abnormal heat generation, and internal short circuit. In [53], a model-based switching model method was proposed to improve the accuracy of the internal resistance estimates during an internal short circuit fault, and RLS was also used for parameter estimation and fault detection. Gao et al. [54] proposed a fault detection method for the micro-short circuit, using the cell difference model and extended Kalman filters to estimate the cell SOC differences. The extra depleting current is identified with RLS to diagnose the short circuit resistance. All the above methods for internal short circuit detection use RLS for parameter estimation for low computational cost, but they all require information from other cells in the pack, which can be influenced by online cell balancing leading to unreliable fault detection.

A string of studies on thermal fault detection using the battery thermal model and the ECM was introduced by the same group of authors in [55–57]. In [55], the Li-ion battery was modeled via ECM and a two-state thermal model. A diagnostic scheme based on the Luenberger observer was proposed to detect and isolate three thermal faults, which were internal thermal resistance fault, convective cooling resistance fault, and thermal runaway, by generating and filtering two unique residuals from the models. This work is further developed in [56], where nonlinear observers using Lyapunov analysis was implemented, and an adaptive threshold generator was developed to deal with the presence of modeling uncertainties. Most recently, the authors proposed a partial differential equation model-based scheme with two state observers to improve the accuracy and reliability of thermal fault detection [57]. The method proposed in these works becomes progressively more precise and reliable, but the computational effort also increases as the method develops.

4.1.2. Non-Model-Based Methods

Non-model-based methods include signal processing and knowledge-based methods. These methods primarily rely on battery data collection, although still using battery modeling to an extent. They can improve fault diagnostic accuracy but might require a large amount of fault data, which is often not available, or have very high computational cost, which is impractical for usage in the BMS.

Several recent works implement signal processing for Li-ion battery fault diagnosis. Kong et al. [58] obtained the remaining charging capacity, which can increase due to extra charge depletion caused by micro-short circuiting, from charging cell voltage curve transformation. The micro-short circuit fault was then identified from the remaining charging capacities between adjacent charges, which was used to obtain the leakage current of the shorted cell. The correlation coefficient was used in [59] to detect the initial stage of short circuits by capturing the off-trend voltage drop and reflect the drop variation in the correlation coefficient. Another work [60] proposed the use of an interclass correlation coefficient to analyze battery short circuit fault also by capturing the off-trend voltage drop. Both methods using the correlation coefficient employ moving windows to prolong the fault memory and do not need extra hardware design or model adjustment. However, they can be subject to measurement noises. Entropy theory fault diagnosis has also been studied recently. Authors in [61] implemented the Shannon entropy and the Z-score method to detect any abnormality in the battery temperature, as well as predicting the time and location of the fault, to prevent thermal runaway. Liu et al. [62,63] proposed the use of a modified Shannon entropy with the Z-score method to capture abnormality in cell voltage, and predict the time and location of the voltage fault occurrence. The entropy-based methods are effective in detecting battery faults, but the computational cost increases with the desired diagnostic precision.

Knowledge-based is another type of non-model-based fault diagnostic method. In some earlier works, Xiong et al. [64] implemented an expert system, including a rule-based method and a probabilistic method, to detect an overdischarge fault, where the rules for failure are an unusual increase in temperature and unusual decrease in voltage. This method is easy to implement, but the threshold for the rules is very difficult to obtain for different applications and cell chemistries. In [65], the authors characterized external short circuit faults through three criteria, which are voltage variation, current variation, and temperature change rate, and constructed thresholds for these criteria to detect external short circuit faults. This method is simple to implement, but it is unreliable as Li-ion cells vary in characteristics, and more experiments would need to be conducted to validate the set thresholds. In [66], the authors used fuzzy logic to analyze temperature, SOC, and voltage residuals from an electrochemical model to detect overcharge, overdischarge, and accelerated degradation. Later works explore the application of machine learning in Li-ion battery fault diagnosis. The Random forests classifier was used in [67] to detect the electrolyte leakage behavior that occurs during an external short circuit fault. The two root-mean-square-error indicators for an external short circuit fault in this study are higher maximum temperature rise and lower discharge capacity. The result of the leakage condition is obtained through training a large amount of data using the classifier. The normal and faulty cells can then be classified effectively, and external short circuit fault can be diagnosed accurately. Zhao et al. [68] established a battery fault diagnostic model through the combination of the 3σ multi-level screening strategy and the machine learning algorithm. The 3σ multi-level screening strategy was utilized to build the criteria for normal operating cell voltage, and a neural network was applied to simulate the cell fault distribution in a battery pack. This method requires an extended period to collect battery data to detect battery faults reliably. Djeziri et al. [69] proposed the use of a Wiener process to model and monitor the drift of voltage to obtain the voltage identification (VID), which determines the voltage needed for a processor to run continuously and at 100%. This data-driven method can be used to predict the degradation trajectory of a system and adjust the parameters accordingly. Although it has not been tested with a Li-ion battery system, its ability to adapt to the degradation process shows its potential to be effective in Li-ion battery fault diagnosis.

4.2. External Battery Fault Diagnosis

4.2.1. Model-Based Methods

For external battery fault diagnosis, model-based methods are often implemented to detect sensor faults and cooling system faults. In some earlier works, Marcicki et al. [70] proposed the use of nonlinear parity equations to generate residuals for voltage, current, and fan setting from the ECM, thermal, and SOC models. The thresholds for the residuals were selected using the probability density function. Current and voltage sensor faults, as well as cooling fan fault, were then detected and isolated. However, this method can only detect faults of large magnitude due to errors from the observer. In [71], sensor fault detection was achieved using the Kirchhoff circuits' equations to generate current and voltage residuals and the temperature diffusion model to generate temperature residuals. This method has a low computational cost but is sensitive to measurement noises. In more recent works, Xu et al. [72] used the proportional-integral-observer-based method to accurately detect and estimate current sensor fault, as a means to improve the state of energy estimation. In [73], a model-based diagnostic scheme using sliding mode observers designed based on the electrical and thermal dynamics of the battery was developed. A set of fault detection filter expressions was derived on the sliding surfaces of each observer. The outputs of these filters were used as residual signals to detect, isolate, and estimate voltage, current and temperature sensor faults, under the assumption that the faults and their time derivatives are bounded and finite. Tran et al. [35] proposed a parameter estimation method using RLS to detect and isolate voltage and current sensor faults. The ECM was used, and the estimated parameters were put through a weighted moving average filter and a statistical cumulative sum (CUSUM) test to evaluate the generated residuals for any abnormalities. This method was also able to detect sensor faults when the battery underwent degradation accurately. The authors in [36] presented a fault diagnostic method for current and voltage sensor faults using state estimation, specifically SOC estimation. The residual was generated from the true SOC, calculated by the coulomb counting method, and the estimated SOC, obtained by the recursive least squares and unscented Kalman filter joint estimation method. Due to sensor noises in real applications, there are some uncertainties associated with the threshold selection for the residuals in this method, which require further research.

A group of authors produced a series of studies on sensor fault diagnosis using structural analysis [2,34,74] and an extended Kalman filter [75–77]. In [34] and [74], structural analysis was applied to obtain the structurally overdetermined part of the system model by analyzing the structural model represented by an incidence matrix. Sequential residual generation was then used as a diagnostic test to detect different faults, including voltage and current sensor faults and cooling system faults, by calculating the unknown variables in a sequence. The specific fault was isolated with many different minimal structurally overdetermined sets. This method was improved in [2] by applying an extended Kalman filter to address the issue of inaccurate initial SOC. In addition, a statistical CUSUM test was implemented to evaluate the generated residuals from the diagnostic tests to determine the presence of a fault. This addition, however, made the proposed method significantly more computationally complex. Another model-based fault diagnostic method was developed by the same authors using an extended Kalman filter in [75]. The extended Kalman filter was used to estimate the output voltage from the ECM to generate residuals between measured and estimated values. The residuals were then evaluated by a statistical CUSUM test to detect current and voltage sensor faults. This work was further developed in [76] and [77] by implementing an adaptive extended Kalman filter to update the process and measurement noise matrices appropriately, which helps eliminate the noises.

4.2.2. Non-Model-Based Methods

For cell connection fault diagnosis, entropy theory and statistical analysis methods have been proposed. Zheng et al. [78] used demonstration data from 96 cells in series collected over three months to obtain cell resistances from the ECM. By calculating the Shannon entropy of the cell resistances, the contact resistance fault was successfully isolated. In [39], the authors proposed a method of fault detection for the connection of Li-ion cells based on entropy. Cell voltage data was obtained and filtered using the discrete cosine filter method. Ensemble Shannon entropy, local Shannon entropy, and sample entropy were applied to predict the time and location of the connection failure occurrence based on voltage fluctuations. Sample entropy produced the most accurate results but required a large amount of data, while ensemble Shannon entropy was able to predict the connection fault between cells effectively. Sun et al. [79] used the wavelet decomposition with three layers developed by Daubechies to smooth the voltage signal and eliminate noise interference. After the wavelet transformation, the Shannon entropy of charge and discharge cycles was calculated, and the change in entropy was monitored to detect real-time cell connection fault. However, this method can be easily affected by the set value of the interval parameters. Ma et al. [80] proposed the use of a modified Z-score test to analyze the abnormal voltage coefficients to detect cell connection fault. The temperature rise rate was utilized as a secondary parameter to determine the severity of the fault. However, using this method, a fault occurring at a location where there is no cross-voltage test cannot be reliably detected.

Since battery packs are often made up of many cells in series, topology-based fault diagnostic methods have been proposed by a few studies. Xia et al. [81] proposed a fault-tolerant voltage measurement method, where the voltage sum of multiple cells was measured instead of the voltage of individual cells. A matrix interpretation of the sensor topology was developed to isolate sensor and cell faults by locating abnormal signals. Since this method is sensitive to measurement noises, the authors further developed an improved measurement topology in [82], where the noise limit and trend of the interleaved voltage measurement method were derived to improve the noise sensitivity. In [83], a multi-fault diagnostic strategy based on an interleaved voltage measurement topology and an improved correlation coefficient method was presented, which can diagnose several types of faults, such as internal and external short circuits, sensor faults and connection faults. The voltage measurement method correlated each battery and contact resistance with two different sensors to accurately identify the location and type of faults. The improved correlation coefficient method was used to monitor fault signatures to eliminate the effect of battery inconsistencies and measurement errors. These topology-based methods can isolate various faults accurately, but they are only suitable for battery packs that have multiple cells in series with interleaved voltage measurements.

4.3. Current Progress and Future Challenges of Li-Ion Battery Fault Diagnosis

In summary, the fault diagnostic algorithms that were discussed have made certain progress on improving Li-ion battery safety, but they still have some limitations in real-life applications. A summary of all the reviewed algorithms is shown in Table 1. Model-based methods can quickly detect and isolate a fault in real-time but require high modeling accuracy. Therefore, further research needs to be conducted to reach a better understanding of the internal battery operation to develop a precise, but not overly complicated, battery model. In addition, the trade-off between robustness and sensitivity of the diagnostic approach, which depends on the fault thresholds, should also be considered. Non-model-based methods can avoid the difficult requirements for battery modeling but still have some drawbacks. Signal processing methods have good dynamic performance but are sensitive to measurement noises and are not able to detect early faults reliably. Simple knowledge-based methods, such as expert systems, require effective rules to precisely detect faults, which is challenging as some battery faults are still not fully understood. Complex knowledge-based methods, such as machine learning, have high accuracy and compatibility with a nonlinear system, such as the Li-ion battery, but the training process is time-consuming and requires a large amount of data. Overall, even though model-based methods are more common, many difficulties still exist in improving battery model accuracy, especially throughout the entire lifespan of the battery. Non-model-based methods, particularly data-driven methods, can have a crucial role in predicting battery behavior as it degrades and aiding the model development process. Therefore, the most effective approach for Li-ion battery fault diagnosis should be a combination of both model-based and non-model-based methods.

There are several challenges that all of these fault diagnostic methods face. First, because some faults have similar effects on the battery, it is difficult to isolate each fault accurately to provide appropriate responses. Current methods often assume that the other components in the system are operating normally to avoid the need to isolate and identify the detected faults. Future research should focus on the development of fault identification methods after successful fault detection. Another challenging task is the determination of effective fault thresholds for early and accurate detection, due to the lack of understanding of fault behavior. Furthermore, a better understanding of battery fault behavior is needed to develop fault simulation tools, because producing a real physical fault can be impractical, costly, and unsafe. Therefore, more studies on the behavior of faults need to be conducted through well-designed experiments to collect data for modeling and simulation purposes. Finally, the BMS computational capability needs to be enhanced to accommodate more complex algorithms that can improve fault diagnostic accuracy significantly. This can be achieved by future hardware development in the BMS, as well as the utilization of cloud-based technologies for battery condition monitoring.

Table 1. Summary of Lithium-ion (Li-ion) fault diagnostic algorithms.

Algorithm Types	Definitions	Algorithms	References
State estimation	The system state is estimated from a model using filters or observers. A fault is detected from the residuals between estimated and measured values.	Particle filter	[46]
		Kalman filter	[2,36,47,48,52,54,75–77]
		Luenberger observer	[49,55]
		Lyapunov-analysis-based nonlinear observer	[56]
		Partial-differential- equation-based observer	[57]
		Proportional integral observer	[72]
		Sliding mode observer	[73]
Parameter estimation	The model parameter is estimated from the measurements using filter algorithms. A fault is detected from the change in the estimated model parameter.	Recursive least squares	[35,36,50–54]
Parity space	A fault is detected through generating residuals from the input and output relationship between the model and the measurements.	Nonlinear parity equations	[70,71]
Structural analysis	The structural overdetermined part of the system model is analyzed to detect and isolate a fault.	Structural analysis	[2,34,74]
Signal processing	Measured signals are transformed into fault parameters, such as entropy or correlation coefficient. A fault is detected from abnormalities in these fault parameters.	Wavelet transform	[58]
		Correlation coefficient	[59,60,83]
		Shannon entropy	[39,61–63,78–80]
		Sensor topology	[81–83]
Knowledge-based	These algorithms use the knowledge obtained from observations or data coming from the system to establish rules or train data to detect a fault.	Rule-based	[64,65]
		Fuzzy logic	[66]
		Random forests classifier	[67]
		Neural network	[68]

5. Conclusions

The safety of the Li-ion battery system has attracted a considerable amount of attention from researchers. Battery faults, including internal and external faults, can hinder the operation of the battery and lead to many potentially hazardous consequences, including fires or explosion. One main function of the BMS is fault diagnosis, which is responsible for detecting faults early and providing control actions to minimize fault effects. Therefore, Li-ion battery fault diagnostic methods have been extensively developed in recent years. This paper provides a comprehensive review of existing fault diagnostic methods for the Li-ion battery system.

Fault diagnostic approaches are categorized into model-based and non-model-based methods. Model-based methods often have low computational cost and fast detection time but are reliant on the accuracy of battery modeling. A simple and precise battery model has not been fully developed. Non-model-based methods are less reliant on battery modeling. However, they require a time-consuming training process that needs a large amount of data. There has not been an effective and practical solution to detect and isolate all potential faults in the Li-ion battery system. There are several challenges in Li-ion battery fault diagnosis, including assumption-free fault isolation, fault threshold selection, fault simulation tools development, and BMS hardware limitations. The summary of the algorithms provided in this paper serves as a basis for researchers to develop more effective fault diagnostic methods for Li-ion battery systems in the future to improve battery safety.

Author Contributions: Conceptualization, M.-K.T. and M.F.; methodology, M.F.; investigation, M.-K.T.; resources, M.F.; writing—original draft preparation, M.-K.T.; writing—review and editing, M.-K.T. and M.F.; visualization, M.-K.T.; supervision, M.F.; project administration, M.-K.T. All authors have read and agreed to the published version of the manuscript.

Funding: This research received no external funding.

Acknowledgments: This work was supported by equipment and manpower from the Department of Chemical Engineering at the University of Waterloo. Special thanks to Danielle Skeba and Priyanka Jindal for their contribution in editing the paper.

Conflicts of Interest: The authors declare no conflict of interest.

References

1. Wu, C.; Zhu, C.; Ge, Y.; Zhao, Y. A Review on Fault Mechanism and Diagnosis Approach for Li-Ion Batteries. *J. Nanomater.* **2015**, *2015*, 1–9. [CrossRef]
2. Liu, Z.; Ahmed, Q.; Zhang, J.; Rizzoni, G.; He, H. Structural analysis based sensors fault detection and isolation of cylindrical lithium-ion batteries in automotive applications. *Control Eng. Pract.* **2016**, *52*, 46–58. [CrossRef]
3. Liu, K.; Liu, Y.; Lin, D.; Pei, A.; Cui, Y. Materials for lithium-ion battery safety. *Sci. Adv.* **2018**, *4*, eaas9820. [CrossRef] [PubMed]
4. Kong, L.; Li, C.; Jiang, J.; Pecht, M. Li-Ion Battery Fire Hazards and Safety Strategies. *Energies* **2018**, *11*, 2191. [CrossRef]
5. Wei, J.; Dong, G.; Chen, Z. Model-based fault diagnosis of Lithium-ion battery using strong tracking Extended Kalman Filter. *Energy Procedia* **2019**, *158*, 2500–2505. [CrossRef]
6. Kim, H.; Shin, K.G. Modeling of externally-induced/common-cause faults in fault-tolerant systems. In Proceedings of the AIAA/IEEE Digital Avionics Systems Conference. 13th DASC, Phoenix, AZ, USA, 30 October–3 November 1994; pp. 402–407.
7. Doughty, D.; Roth, E.P. A General Discussion of Li Ion Battery Safety. *Electrochem. Soc. Interface* **2012**, *21*, 37–44.
8. Lu, L.; Han, X.; Li, J.; Hua, J.; Ouyang, M. A review on the key issues for lithium-ion battery management in electric vehicles. *J. Power Sources* **2013**, *226*, 272–288. [CrossRef]
9. Venkatasubramanian, V.; Rengaswamy, R.; Yin, K.; Kavuri, S.N. A review of process fault detection and diagnosis. *Comput. Chem. Eng.* **2003**, *27*, 293–311. [CrossRef]

10. Xiong, R.; Yu, Q.; Shen, W. Review on sensors fault diagnosis and fault-tolerant techniques for lithium ion batteries in electric vehicles. In Proceedings of the 2018 13th IEEE Conference on Industrial Electronics and Applications (ICIEA), Wuhan, China, 31 May–2 June 2018; pp. 406–410.

11. Lyu, D.; Ren, B.; Li, S. Failure modes and mechanisms for rechargeable Lithium-based batteries: A state-of-the-art review. *Acta Mech.* **2019**, *230*, 701–727. [CrossRef]

12. Feng, X.; Sun, J.; Ouyang, M.; Wang, F.; He, X.; Lu, L.; Peng, H. Characterization of penetration induced thermal runaway propagation process within a large format lithium ion battery module. *J. Power Sources* **2015**, *275*, 261–273. [CrossRef]

13. Lelie, M.; Braun, T.; Knips, M.; Nordmann, H.; Ringbeck, F.; Zappen, H.; Sauer, D. Battery Management System Hardware Concepts: An Overview. *Appl. Sci.* **2018**, *8*, 534. [CrossRef]

14. Liu, Y.; Xie, J. Failure Study of Commercial LiFePO 4 Cells in Overcharge Conditions Using Electrochemical Impedance Spectroscopy. *J. Electrochem. Soc.* **2015**, *162*, A2208–A2217. [CrossRef]

15. Larsson, F.; Andersson, P.; Blomqvist, P.; Mellander, B.-E. Toxic fluoride gas emissions from lithium-ion battery fires. *Sci. Rep.* **2017**, *7*, 10018. [CrossRef] [PubMed]

16. Ouyang, D.; Chen, M.; Liu, J.; Wei, R.; Weng, J.; Wang, J. Investigation of a commercial lithium-ion battery under overcharge/over-discharge failure conditions. *RSC Adv.* **2018**, *8*, 33414–33424. [CrossRef]

17. Guo, R.; Lu, L.; Ouyang, M.; Feng, X. Mechanism of the entire overdischarge process and overdischarge-induced internal short circuit in lithium-ion batteries. *Sci. Rep.* **2016**, *6*, 30248. [CrossRef]

18. Fear, C.; Juarez-Robles, D.; Jeevarajan, J.A.; Mukherjee, P.P. Elucidating Copper Dissolution Phenomenon in Li-Ion Cells under Overdischarge Extremes. *J. Electrochem. Soc.* **2018**, *165*, A1639–A1647. [CrossRef]

19. Wang, H.; Simunovic, S.; Maleki, H.; Howard, J.N.; Hallmark, J.A. Internal configuration of prismatic lithium-ion cells at the onset of mechanically induced short circuit. *J. Power Sources* **2016**, *306*, 424–430. [CrossRef]

20. Abaza, A.; Ferrari, S.; Wong, H.K.; Lyness, C.; Moore, A.; Weaving, J.; Blanco-Martin, M.; Dashwood, R.; Bhagat, R. Experimental study of internal and external short circuits of commercial automotive pouch lithium-ion cells. *J. Energy Storage* **2018**, *16*, 211–217. [CrossRef]

21. Mao, B.; Chen, H.; Cui, Z.; Wu, T.; Wang, Q. Failure mechanism of the lithium ion battery during nail penetration. *Int. J. Heat Mass Transf.* **2018**, *122*, 1103–1115. [CrossRef]

22. Kriston, A.; Pfrang, A.; Döring, H.; Fritsch, B.; Ruiz, V.; Adanouj, I.; Kosmidou, T.; Ungeheuer, J.; Boon-Brett, L. External short circuit performance of Graphite-LiNi1/3Co1/3Mn1/3O2 and Graphite-LiNi0.8Co0.15Al0.05O2 cells at different external resistances. *J. Power Sources* **2017**, *361*, 170–181. [CrossRef]

23. Rheinfeld, A.; Sturm, J.; Frank, A.; Kosch, S.; Erhard, S.V.; Jossen, A. Impact of Cell Size and Format on External Short Circuit Behavior of Lithium-Ion Cells at Varying Cooling Conditions: Modeling and Simulation. *J. Electrochem. Soc.* **2020**, *167*, 13511. [CrossRef]

24. Panda, S.; Sahu, B.K.; Mohanty, P.K. Design and performance analysis of PID controller for an automatic voltage regulator system using simplified particle swarm optimization. *J. Frankl. Inst.* **2012**, *349*, 2609–2625. [CrossRef]

25. Lystianingrum, V.; Hredzak, B.; Agelidis, V.G. Multiple model estimator based detection of abnormal cell overheating in a Li-ion battery string with minimum number of temperature sensors. *J. Power Sources* **2015**, *273*, 1171–1181. [CrossRef]

26. Ruiz, V.; Pfrang, A.; Kriston, A.; Omar, N.; van den Bossche, P.; Boon-Brett, L. A review of international abuse testing standards and regulations for lithium ion batteries in electric and hybrid electric vehicles. *Renew. Sustain. Energy Rev.* **2018**, *81*, 1427–1452. [CrossRef]

27. Diao, W.; Xing, Y.; Saxena, S.; Pecht, M. Evaluation of Present Accelerated Temperature Testing and Modeling of Batteries. *Appl. Sci.* **2018**, *8*, 1786. [CrossRef]

28. Xu, B.; Shi, Y.; Kirschen, D.S.; Zhang, B. Optimal regulation response of batteries under cycle aging mechanisms. In Proceedings of the 2017 IEEE 56th Annual Conference on Decision and Control (CDC), Melbourne, Australia, 12–15 December 2017; pp. 751–756.

29. Xu, J.; Deshpande, R.D.; Pan, J.; Cheng, Y.-T.; Battaglia, V.S. Electrode Side Reactions, Capacity Loss and Mechanical Degradation in Lithium-Ion Batteries. *J. Electrochem. Soc.* **2015**, *162*, A2026–A2035. [CrossRef]

30. Kanevskii, L.S.; Dubasova, V.S. Degradation of Lithium-Ion batteries and how to fight it: A review. *Russ. J. Electrochem.* **2005**, *41*, 1–16. [CrossRef]

31. Ma, S.; Jiang, M.; Tao, P.; Song, C.; Wu, J.; Wang, J.; Deng, T.; Shang, W. Temperature effect and thermal impact in lithium-ion batteries: A review. *Prog. Nat. Sci. Mater. Int.* **2018**, *28*, 653–666. [CrossRef]
32. Wilke, S.; Schweitzer, B.; Khateeb, S.; Al-Hallaj, S. Preventing thermal runaway propagation in lithium ion battery packs using a phase change composite material: An experimental study. *J. Power Sources* **2017**, *340*, 51–59. [CrossRef]
33. Galushkin, N.E.; Yazvinskaya, N.N.; Galushkin, D.N. Mechanism of Thermal Runaway in Lithium-Ion Cells. *J. Electrochem. Soc.* **2018**, *165*, A1303–A1308. [CrossRef]
34. Liu, Z.; Ahmed, Q.; Rizzoni, G.; He, H. Fault Detection and Isolation for Lithium Ion Battery System Using Structural Analysis and Sequential Residual Generation. In Proceedings of the ASME 7th annual dynamic systems and control conference 2014, San Antonio, TX, USA, 22–24 October 2014.
35. Tran, M.-K.; Fowler, M. Sensor Fault Detection and Isolation for Degrading Lithium-Ion Batteries in Electric Vehicles Using Parameter Estimation with Recursive Least Squares. *Batteries* **2020**, *6*, 1. [CrossRef]
36. Xiong, R.; Yu, Q.; Shen, W.; Lin, C.; Sun, F. A Sensor Fault Diagnosis Method for a Lithium-Ion Battery Pack in Electric Vehicles. *IEEE Trans. Power Electron.* **2019**, *34*, 9709–9718. [CrossRef]
37. Zheng, C.; Chen, Z.; Huang, D. Fault diagnosis of voltage sensor and current sensor for lithium-ion battery pack using hybrid system modeling and unscented particle filter. *Energy* **2020**, *191*, 116504. [CrossRef]
38. Xia, G.; Cao, L.; Bi, G. A review on battery thermal management in electric vehicle application. *J. Power Sources* **2017**, *367*, 90–105. [CrossRef]
39. Yao, L.; Wang, Z.; Ma, J. Fault detection of the connection of lithium-ion power batteries based on entropy for electric vehicles. *J. Power Sources* **2015**, *293*, 548–561. [CrossRef]
40. Offer, G.J.; Yufit, V.; Howey, D.A.; Wu, B.; Brandon, N.P. Module design and fault diagnosis in electric vehicle batteries. *J. Power Sources* **2012**, *206*, 383–392. [CrossRef]
41. Rahimi-Eichi, H.; Ojha, U.; Baronti, F.; Chow, M.-Y. Battery Management System: An Overview of Its Application in the Smart Grid and Electric Vehicles. *EEE Ind. Electron. Mag.* **2013**, *7*, 4–16. [CrossRef]
42. Brand, M.; Glaser, S.; Geder, J.; Menacher, S.; Obpacher, S.; Jossen, A.; Quinger, D. Electrical safety of commercial Li-ion cells based on NMC and NCA technology compared to LFP technology. In Proceedings of the 2013 World Electric Vehicle Symposium and Exhibition (EVS27), Barcelona, Spain, 17–20 November 2013; pp. 1–9.
43. Hendricks, C.; Williard, N.; Mathew, S.; Pecht, M. A failure modes, mechanisms, and effects analysis (FMMEA) of lithium-ion batteries. *J. Power Sources* **2015**, *297*, 113–120. [CrossRef]
44. Alavi, S.M.M.; Samadi, M.F.; Saif, M. Diagnostics in Lithium-Ion Batteries: Challenging Issues and Recent Achievements. In *Integration of Practice-Oriented Knowledge Technology: Trends and Prospectives*; Fathi, M., Ed.; Springer: Berlin/Heidelberg, Germany, 2013; pp. 277–291.
45. Tomasov, M.; Kajanova, M.; Bracinik, P.; Motyka, D. Overview of Battery Models for Sustainable Power and Transport Applications. *Transp. Res. Procedia* **2019**, *40*, 548–555. [CrossRef]
46. Alavi, S.M.M.; Samadi, M.F.; Saif, M. Plating Mechanism Detection in Lithium-ion batteries, by using a particle-filtering based estimation technique. In Proceedings of the 2013 American Control Conference, Washington, DC, USA, 17–19 June 2013; pp. 4356–4361.
47. Singh, A.; Izadian, A.; Anwar, S. Fault diagnosis of Li-Ion batteries using multiple-model adaptive estimation. In Proceedings of the IECON 2013, 39th annual conference of the IEEE Industrial Electronics Societ, Vienna, Austria, 10–13 November 2013.
48. Sidhu, A.; Izadian, A.; Anwar, S. Adaptive Nonlinear Model-Based Fault Diagnosis of Li-Ion Batteries. *IEEE Trans. Ind. Electron.* **2015**, *62*, 1002–1011. [CrossRef]
49. Chen, W.; Chen, W.-T.; Saif, M.; Li, M.-F.; Wu, H. Simultaneous Fault Isolation and Estimation of Lithium-Ion Batteries via Synthesized Design of Luenberger and Learning Observers. *IEEE Trans. Contr. Syst. Technol.* **2014**, *22*, 290–298. [CrossRef]
50. Ouyang, M.; Zhang, M.; Feng, X.; Lu, L.; Li, J.; He, X.; Zheng, Y. Internal short circuit detection for battery pack using equivalent parameter and consistency method. *J. Power Sources* **2015**, *294*, 272–283. [CrossRef]
51. Feng, X.; Weng, C.; Ouyang, M.; Sun, J. Online internal short circuit detection for a large format lithium ion battery. *Appl. Energy* **2016**, *161*, 168–180. [CrossRef]
52. Feng, X.; Pan, Y.; He, X.; Wang, L.; Ouyang, M. Detecting the internal short circuit in large-format lithium-ion battery using model-based fault-diagnosis algorithm. *J. Energy Storage* **2018**, *18*, 26–39. [CrossRef]

53. Seo, M.; Goh, T.; Park, M.; Koo, G.; Kim, S. Detection of Internal Short Circuit in Lithium Ion Battery Using Model-Based Switching Model Method. *Energies* **2017**, *10*, 76. [CrossRef]
54. Gao, W.; Zheng, Y.; Ouyang, M.; Li, J.; Lai, X.; Hu, X. Micro-Short-Circuit Diagnosis for Series-Connected Lithium-Ion Battery Packs Using Mean-Difference Model. *IEEE Trans. Ind. Electron.* **2019**, *66*, 2132–2142. [CrossRef]
55. Dey, S.; Biron, Z.A.; Tatipamula, S.; Das, N.; Mohon, S.; Ayalew, B.; Pisu, P. On-board Thermal Fault Diagnosis of Lithium-ion Batteries For Hybrid Electric Vehicle Application. *IFAC PapersOnLine* **2015**, *48*, 389–394. [CrossRef]
56. Dey, S.; Biron, Z.A.; Tatipamula, S.; Das, N.; Mohon, S.; Ayalew, B.; Pisu, P. Model-based real-time thermal fault diagnosis of Lithium-ion batteries. *Control Eng. Pract.* **2016**, *56*, 37–48. [CrossRef]
57. Dey, S.; Perez, H.E.; Moura, S.J. Model-Based Battery Thermal Fault Diagnostics: Algorithms, Analysis, and Experiments. *IEEE Trans. Contr. Syst. Technol.* **2019**, *27*, 576–587. [CrossRef]
58. Kong, X.; Zheng, Y.; Ouyang, M.; Lu, L.; Li, J.; Zhang, Z. Fault diagnosis and quantitative analysis of micro-short circuits for lithium-ion batteries in battery packs. *J. Power Sources* **2018**, *395*, 358–368. [CrossRef]
59. Xia, B.; Shang, Y.; Nguyen, T.; Mi, C. A correlation based fault detection method for short circuits in battery packs. *J. Power Sources* **2017**, *337*, 1–10. [CrossRef]
60. Li, X.; Wang, Z. A novel fault diagnosis method for lithium-Ion battery packs of electric vehicles. *Measurement* **2018**, *116*, 402–411. [CrossRef]
61. Hong, J.; Wang, Z.; Liu, P. Big-Data-Based Thermal Runaway Prognosis of Battery Systems for Electric Vehicles. *Energies* **2017**, *10*, 919. [CrossRef]
62. Wang, Z.; Hong, J.; Liu, P.; Zhang, L. Voltage fault diagnosis and prognosis of battery systems based on entropy and Z -score for electric vehicles. *Appl. Energy* **2017**, *196*, 289–302. [CrossRef]
63. Liu, P.; Sun, Z.; Wang, Z.; Zhang, J. Entropy-Based Voltage Fault Diagnosis of Battery Systems for Electric Vehicles. *Energies* **2018**, *11*, 136. [CrossRef]
64. Xiong, J.; Banvait, H.; Li, L.; Chen, Y.; Xie, J.; Liu, Y.; Wu, M.; Chen, J. Failure detection for over-discharged Li-ion batteries. In Proceedings of the 2012 IEEE International Electric Vehicle Conference, Greenville, SC, USA, 4–8 March 2012.
65. Xia, B.; Chen, Z.; Mi, C.; Robert, B. External short circuit fault diagnosis for lithium-ion batteries. In Proceedings of the IEEE Transportation Electrification Conference and Expo (ITEC), Dearborn, MI, USA, 15–18 June 2014.
66. Muddappa, V.K.S.; Anwar, S. Electrochemical Model Based Fault Diagnosis of Li-Ion Battery Using Fuzzy Logic. In Proceedings of the ASME International Mechanical Engineering Congress and Exposition, Montreal, QC, Canada, 14–20 November 2014.
67. Yang, R.; Xiong, R.; He, H.; Chen, Z. A fractional-order model-based battery external short circuit fault diagnosis approach for all-climate electric vehicles application. *J. Clean. Prod.* **2018**, *187*, 950–959. [CrossRef]
68. Zhao, Y.; Liu, P.; Wang, Z.; Zhang, L.; Hong, J. Fault and defect diagnosis of battery for electric vehicles based on big data analysis methods. *Appl. Energy* **2017**, *207*, 354–362. [CrossRef]
69. Djeziri, M.A.; Benmoussa, S.; Benbouzid, M.E. Data-driven approach augmented in simulation for robust fault prognosis. *Eng. Appl. Artif. Intell.* **2019**, *86*, 154–164. [CrossRef]
70. Marcicki, J.; Onori, S.; Rizzoni, G. Nonlinear Fault Detection and Isolation for a Lithium-Ion Battery Management System. In Proceedings of the ASME Dynamic Systems and Control Conference, Cambridge, MA, USA, 12–15 September 2010.
71. Lombardi, W.; Zarudniev, M.; Lesecq, S.; Bacquet, S. Sensors fault diagnosis for a BMS. In Proceedings of the Control Conference (ECC), Strasbourg, France, 24–27 June 2014.
72. Xu, J.; Wang, J.; Li, S.; Cao, B. A Method to Simultaneously Detect the Current Sensor Fault and Estimate the State of Energy for Batteries in Electric Vehicles. *Sensors* **2016**, *16*, 1328. [CrossRef]
73. Dey, S.; Mohon, S.; Pisu, P.; Ayalew, B. Sensor Fault Detection, Isolation, and Estimation in Lithium-Ion Batteries. *IEEE Trans. Contr. Syst. Technol.* **2016**, *24*, 2141–2149. [CrossRef]
74. Liu, Z.; He, H.; Ahmed, Q.; Rizzoni, G. Structural Analysis Based Fault Detection and Isolation Applied for A Lithium-Ion Battery Pack. *IFAC PapersOnLine* **2015**, *48*, 1465–1470. [CrossRef]
75. Liu, Z.; He, H. Model-based Sensor Fault Diagnosis of a Lithium-ion Battery in Electric Vehicles. *Energies* **2015**, *8*, 6509–6527. [CrossRef]

76. He, H.; Liu, Z.; Hua, Y. Adaptive Extended Kalman Filter Based Fault Detection and Isolation for a Lithium-Ion Battery Pack. *Energy Procedia* **2015**, *75*, 1950. [CrossRef]

77. Liu, Z.; He, H. Sensor fault detection and isolation for a lithium-ion battery pack in electric vehicles using adaptive extended Kalman filter. *Appl. Energy* **2017**, *185*, 2033–2044. [CrossRef]

78. Zheng, Y.; Han, X.; Lu, L.; Li, J.; Ouyang, M. Lithium ion battery pack power fade fault identification based on Shannon entropy in electric vehicles. *J. Power Sources* **2013**, *223*, 136–146. [CrossRef]

79. Sun, Z.; Liu, P.; Wang, Z. Real-time Fault Diagnosis Method of Battery System Based on Shannon Entropy. *Energy Procedia* **2017**, *105*, 2354–2359. [CrossRef]

80. Ma, M.; Wang, Y.; Duan, Q.; Wu, T.; Sun, J.; Wang, Q. Fault detection of the connection of lithium-ion power batteries in series for electric vehicles based on statistical analysis. *Energy* **2018**, *164*, 745–756. [CrossRef]

81. Xia, B.; Mi, C. A fault-tolerant voltage measurement method for series connected battery packs. *J. Power Sources* **2016**, *308*, 83–96. [CrossRef]

82. Xia, B.; Nguyen, T.; Yang, J.; Mi, C. The improved interleaved voltage measurement method for series connected battery packs. *J. Power Sources* **2016**, *334*, 12–22. [CrossRef]

83. Kang, Y.; Duan, B.; Zhou, Z.; Shang, Y.; Zhang, C. A multi-fault diagnostic method based on an interleaved voltage measurement topology for series connected battery packs. *J. Power Sources* **2019**, *417*, 132–144. [CrossRef]

MDPI
St. Alban-Anlage 66
4052 Basel
Switzerland
Tel. +41 61 683 77 34
Fax +41 61 302 89 18
www.mdpi.com

Algorithms Editorial Office
E-mail: algorithms@mdpi.com
www.mdpi.com/journal/algorithms

9 783036 504629